Evaluation of Long-Term Water-Level Declines in Basalt Aquifers near Mosier, Oregon

By Erick R. Burns, David S. Morgan, Karl K. Lee, Jonathan V. Haynes, and Terrence D. Conlon

Prepared in cooperation with the Wasco County Soil and Water Conservation District

Scientific Investigations Report 2012–5002

U.S. Department of the Interior
U.S. Geological Survey

U.S. Department of the Interior
KEN SALAZAR, Secretary

U.S. Geological Survey
Marcia K. McNutt, Director

U.S. Geological Survey, Reston, Virginia: 2012

For more information on the USGS—the Federal source for science about the Earth, its natural and living resources, natural hazards, and the environment, visit http://www.usgs.gov or call 1–888–ASK–USGS.

For an overview of USGS information products, including maps, imagery, and publications,
visit http://www.usgs.gov/pubprod

To order this and other USGS information products, visit http://store.usgs.gov

Suggested citation:
Burns, E.R., Morgan, D.S., Lee, K.K., Haynes, J.V., and Conlon, T.D., 2012, Evaluation of long-term water-level declines in basalt aquifers near Mosier, Oregon: U.S. Geological Survey Scientific Investigations Report 2012–5002, 134 p.

Contents

Contents—Continued

Figures

Figures—Continued

Tables

Conversion Factors, Datums, and Abbreviations and Acronyms

Inch/Pound to SI

Multiply	By	To obtain
Length		
inch (in.)	2.54	centimeter (cm)
foot (ft)	0.3048	meter (m)
mile (mi)	1.609	kilometer (km)
Area		
Acre	4,047	square meter (m^2)
square mile (mi^2)	640	acre
square foot (ft^2)	0.09290	square meter (m^2)
Volume		
acre-foot (acre-ft)	43,560	cubic foot (ft^3)
Flow rate		
acre-foot per yr (acre-ft/yr)	119.3	cubic foot per day (ft^3/d)
foot per day (ft/d)	0.3048	meter per day (m/d)
foot per year (ft/yr)	0.3048	meter per year (m/yr)
gallons per minute (gal/min)	192.5	cubic foot per day (ft^3/d)
cubic foot per second (ft^3/s)	86,400	cubic foot per day (ft^3/d)
Transmissivity*		
foot squared per day (ft^2/d)	0.09290	meter squared per day (m^2/d)

SI to Inch/Pound

Multiply	By	To obtain
Length		
centimeter (cm)	0.3937	inch (in.)
meter (m)	3.281	foot (ft)
millimeter (mm)	0.03937	inch (in.)

Temperature in degrees Celsius (°C) may be converted to degrees Fahrenheit (°F) as follows:

$$°F=(1.8 \times °C)+32.$$

*Transmissivity: The standard unit for transmissivity is cubic foot per day per square foot times foot of aquifer thickness [(ft^3/d)/ft^2]ft. In this report, the mathematically reduced form, foot squared per day (ft^2/d), is used for convenience.

Conversion Factors, Datums, and Abbreviations and Acronyms

Datums

Vertical coordinate information is referenced to the North American Vertical Datum of 1988 (NAVD88).

Horizontal coordinate information is referenced to the North American Datum of 1927 (NAD27), utilizing the NAD_1927_StatePlane_Oregon_North_FIPS_3601 projection.

Elevation, as used in this report, refers to distance above the vertical datum in a strict sense when associated with a value and in a broad sense when not associated with a value.

Abbreviations and Acronyms

AR	artificial recharge
ASR	aquifer storage and recovery
CLU	Common Land Unit
CRBG	Columbia River Basalt Group
HFB	Horizontal Flow Barrier
LPF	Layer Property Flow
MRC	master recession curve
OWRD	Oregon Water Resources Department
PART	Stream hydrograph separation computer program that uses the streamflow partitioning method
RM	river mile
RORA	Streamflow recession analysis computer program based on the Rorabaugh method
SWCD	Soil and Water Conservation District
USGS	U.S. Geological Survey
WRIS	Water Rights Information System

Evaluation of Long-Term Water-Level Declines in Basalt Aquifers near Mosier, Oregon

By Erick R. Burns, David S. Morgan, Karl K. Lee, Jonathan V. Haynes, and Terrence D. Conlon

Executive Summary

The Mosier area lies along the Columbia River in northwestern Wasco County between the cities of Hood River and The Dalles, Oregon. Major water uses in the area are irrigation, municipal supply for the city of Mosier, and domestic supply for rural residents. The primary source of water is groundwater from the Columbia River Basalt Group (CRBG) aquifers that underlie the area. Concerns regarding this supply of water arose in the mid-1970s, when groundwater levels in the orchard tract area began to steadily decline. In the 1980s, the Oregon Water Resources Department (OWRD) conducted a study of the aquifer system, which resulted in delineation of an administrative area where parts of the Pomona and Priest Rapids aquifers were withdrawn from further appropriations for any use other than domestic supply. Despite this action, water levels continued to drop at approximately the same, nearly constant annual rate of about 4 feet per year, resulting in a current total decline of between 150 and 200 feet in many wells with continued downward trends.

In 2005, the Mosier Watershed Council and the Wasco Soil and Water Conservation District began a cooperative investigation of the groundwater system with the U.S. Geological Survey. The objectives of the study were to advance the scientific understanding of the hydrology of the basin, to assess the sustainability of the water supply, to evaluate the causes of persistent groundwater-level declines, and to evaluate potential management strategies. An additional U.S. Geological Survey objective was to advance the understanding of CRBG aquifers, which are the primary source of water across a large part of Oregon, Washington, and Idaho. In many areas, significant groundwater level declines have resulted as these aquifers were heavily developed for agricultural, municipal, and domestic water supplies.

Three major factors were identified as possible contributors to the water-level declines in the study area: (1) pumping at rates that are not sustainable, (2) well construction practices that have resulted in leakage from aquifers into springs and streams, and (3) reduction in aquifer recharge resulting from long-term climate variations. Historical well construction practices, specifically open, unlined, uncased boreholes that result in cross-connecting (or commingling) multiple aquifers, allow water to flow between these aquifers. Water flowing along the path of least resistance, through commingled boreholes, allows the drainage of aquifers that previously stored water more efficiently.

The study area is in the eastern foothills of the Cascade Range in north central Oregon in a transitional zone between the High Cascades to the west and the Columbia Plateau to the east. The 78-square mile (mi^2) area is defined by the drainages of three streams—Mosier Creek (51.8 mi^2), Rock Creek (13.9 mi^2), and Rowena Creek (6.9 mi^2)—plus a small area that drains directly to the Columbia River.

The three major components of the study are: (1) a 2-year intensive data collection period to augment previous streamflow and groundwater-level measurements, (2) precipitation-runoff modeling of the watersheds to determine the amount of recharge to the aquifer system, and (3) groundwater-flow modeling and analysis to evaluate the cause of groundwater-level declines and to evaluate possible water resource management strategies.

Data collection included the following:

1. Water-level measurements were made in 37 wells. Bi-monthly or quarterly measurements were made in 30 wells, and continuous water-level monitoring instruments were installed in 7 wells. The measurements principally were made to capture the seasonal patterns in the groundwater system, and to augment the available long-term record.

2. Groundwater pumping was measured, reported, or estimated from irrigation, municipal and domestic wells. Flowmeters were installed on 74 percent of all high-capacity irrigation wells in the study area.

3. Borehole geophysical data were collected from a known commingling well. These data measured geologic properties and vertical flow through the well.

4. Streamflow measurements were made in Rock, Rowena, and Mosier Creeks. A long-term recording stream-gaging station was reestablished on Mosier Creek to provide a continuous record of streamflow. Streamflow measurements also were made along the creeks periodically to evaluate seasonal patterns of exchange between streams and the groundwater system.

Major findings from the study include:

1. Annual average precipitation ranges from 20 to 54 inches across the study area with an average value of about 30 inches. Based on rainfall-runoff modeling, about one-third of this water infiltrates into the aquifer system.

2. Currently, about 3 percent of the water infiltrated into the groundwater system is extracted for municipal, agricultural, and rural residential use. The remainder of the water flows through the aquifer system, discharging into local streams and the Columbia River. About 80 percent of recent pumping supports crop production. The city of Mosier public supply wells account for about 10 percent of total pumping, with the remaining 10 percent being pumped from the private wells of rural residents.

3. Groundwater-flow simulation results indicate that leakage through commingling wells is a significant and likely the dominant cause of water level declines. Leakage patterns can be complex, but most of the leaked water likely flows out the CRBG aquifer system through very permeable sediments into Mosier Creek and its tributary streams in the OWRD administrative area. Model-derived estimates attribute 80-90 percent of the declines to commingling, with pumping accounting for the remaining 10-20 percent. Although decadal trends in precipitation have occurred, associated changes in aquifer recharge are likely not a significant contributor to the current water level declines.

4. As many as 150 wells might be commingling. To evaluate whether or not the local combination of geology and well construction have resulted in aquifer commingling at a particular well, the well needs to be tested by measuring intraborehole flow. During geophysical testing of one known commingling well, the flow rate through the well between aquifers ranged between 70 and 135 gallons per minute (11-22 percent of total annual pumping in the study area). Historically, when aquifer water levels were 150-200 feet higher, this flow rate would have been correspondingly higher.

5. Because aquifer commingling through well boreholes is likely the dominant cause of aquifer declines, flow simulations were conducted to evaluate the benefit of repairing wells in specified locations and the benefit of recharging aquifers using diverted flow from study area creeks. As part of this analysis, maps were generated that show which areas are more vulnerable to commingling. These maps indicate that the value of repairing wells in the area generally coincident with the OWRD administrative area is higher than in areas farther upstream in the watershed. Simulation results also indicate that artificial recharge of the aquifers using diverted creek water will not significantly improve water levels in the aquifer system unless at least some commingling wells are repaired first. Repairs would entail construction of wells in a manner that prevents commingling of multiple aquifers. The value of artificially recharging the aquifers improves as more wells are repaired because the aquifer system more efficiently stores water.

Introduction

The Mosier area lies in northwestern Wasco County between the cities of Hood River and The Dalles, Oregon (fig. 1). Water is needed for irrigation, municipal supply for the city of Mosier, and domestic use for rural residents. The primary source of water is groundwater within the Columbia River Basalt Group (CRBG) aquifers that underlie the area. Concerns regarding the sustainability of using the CRBG aquifers for long-term water supply have grown during the past 30–40 years as water levels in the aquifers have steadily declined.

Groundwater levels began declining in the 1970s during a period of intense development of groundwater resources. Causes for the declines are pumping and leakage between aquifers through well boreholes open to multiple aquifers (commingling wells) (Lite and Grondin, 1988); however, the relative importance of these factors was unknown. Following a hydrogeologic assessment by the Oregon Water Resources Department (OWRD) (Lite and Grondin, 1988), a groundwater administrative area was delineated (fig. 1), and the Pomona and Priest Rapids aquifers in the area were withdrawn from further appropriations for any use other than domestic supply. Since that time, water levels in the area have continued to decline steadily. Among the adverse effects of the groundwater declines are (1) increased energy costs for pumping, (2) expense of deepening or replacing wells, and (3) reduced groundwater discharge to streams that can affect aquatic habitat (Lite and Grondin, 1988). Continued declines can further reduce flow in streams and make it infeasible for groundwater to support current water demand.

The Mosier Watershed Council and Wasco County Soil and Water Conservation District (SWCD) have established three goals for the watershed: (1) to reverse or stabilize water-level declines in the principal aquifers of the Mosier area, (2) to increase summer base flows in Mosier Creek, and (3) to sustain productive, profitable agriculture in Mosier Valley (Jennifer Clark, Mosier Watershed Council, written commun., 2004). To meet these goals, the Mosier Watershed Council and SWCD are working with the OWRD to identify groundwater management strategies to ensure groundwater resources will sustain future water needs.

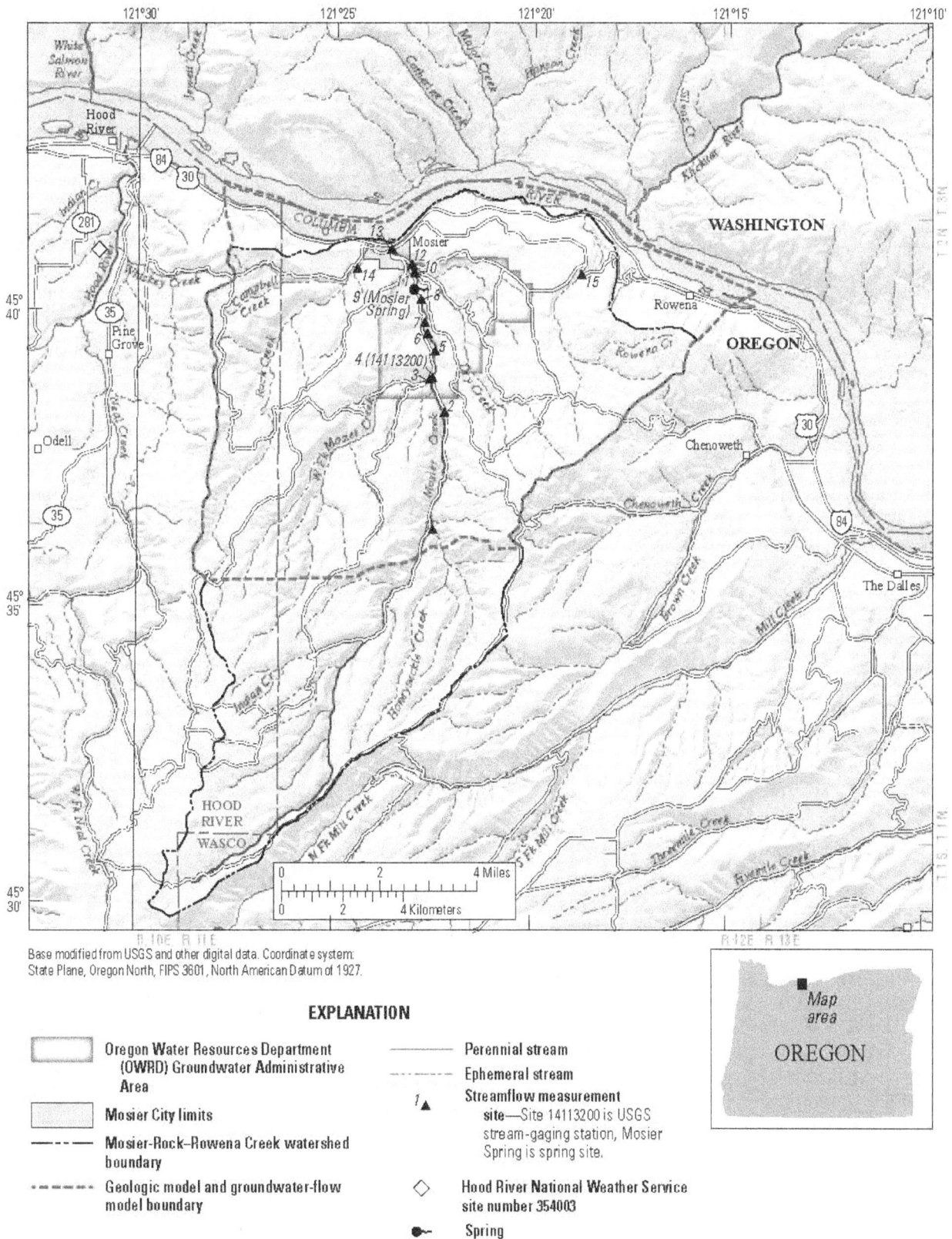

Base modified from USGS and other digital data. Coordinate system:
State Plane, Oregon North, FIPS 3601, North American Datum of 1927.

EXPLANATION

| | Oregon Water Resources Department (OWRD) Groundwater Administrative Area |
| | Mosier City limits |

—·—·— Mosier-Rock–Rowena Creek watershed boundary

- - - - - Geologic model and groundwater-flow model boundary

——— Perennial stream

—·—·— Ephemeral stream

1 ▲ Streamflow measurement site—Site 14113200 is USGS stream-gaging station, Mosier Spring is spring site.

◇ Hood River National Weather Service site number 354003

●— Spring

Figure 1.   Extent of the drainage area covered by the rainfall-runoff simulation model, and extent of the geologic and groundwater-flow simulation models for the Mosier, Oregon, study area.

In 2005, the Mosier Watershed Council and SWCD began a cooperative investigation of the groundwater system with the U.S. Geological Survey (USGS) to advance the scientific understanding of the hydrology of the basin and use that understanding to develop tools that can be used to evaluate management strategies. Another objective of the study was to advance the understanding of CRBG aquifers. These aquifers are some of the most productive aquifers in Oregon, Washington, and Idaho, and in some locations, these aquifers are heavily developed for agricultural, municipal, and domestic water supplies. Many other areas also have experienced significant groundwater-level declines, and water managers are seeking to achieve sustainable levels of groundwater development in the CRBG aquifers.

Purpose and Scope

The purpose of this report is to identify the causes of long-term groundwater-level declines within basalt aquifers in the Mosier area. The first part of this report summarizes the purpose and scope of this study and provides a description of the study area and previous investigations. The second part describes the hydrogeology of the study area including the geologic and hydrogeologic frameworks, important components of the water budget, and groundwater flow. The final part summarizes the development and use of a three-dimensional numerical model of the groundwater-flow system to evaluate the causes of groundwater level declines and forecast the effects of management options. This report has six appendixes containing supplementary material. The technical details of construction of the groundwater-flow simulation model and supporting input data are summarized in appendixes A through E. The results of geophysical testing of well boreholes are summarized in appendix F.

Description of Study Area

The Mosier area is in the eastern foothills of the Cascade Range in north central Oregon in a transitional zone between the High Cascades to the west and the Columbia Plateau to the east (fig. 1). The 78 mi² area is defined by the drainages of three streams—Mosier Creek (51.8 mi²), Rock Creek (13.9 mi²), and Rowena Creek (6.9 mi²)—all of which are tributary to the Columbia River. The area drains to the north with elevations ranging from more than 2,300 ft at Wasco Butte to about 70 ft at the Columbia River. The climate is semi-arid to dry sub-humid. The distribution of precipitation

in the study area follows a strong gradient, decreasing from higher to lower elevations owing to orographic effects from west to east (associated with the Cascade Range) and south to north (associated with change in elevation along the watershed drainage), based on 1971–2000 average precipitation (PRISM Group, 2010). Average annual precipitation in the northwestern part of the study area is about 35 in., decreasing to 16 in. toward the northeast. Average annual precipitation in the southern part of the study area in the headwaters of Mosier Creek is 57 in., compared to 24 in. at the mouth of Mosier Creek. The distribution of ambient temperature in the study area also follows gradients from west to east and south to north. The Columbia River Gorge connects the moderate marine climate to the west with the interior climate to the east. Temperature increases due to orographic effects from the upland area in the south toward the lowland in the north.

Previous Investigations

In the earliest relevant work, a small part of the current study area was covered by Piper's (1932) general description of the geology and hydrogeology of The Dalles area. This description was developed further in papers by Newcomb (1961, 1963, and 1969) describing the occurrence and flow of groundwater through important aquifers in the vicinity, especially the younger Dalles Formation (volcaniclastic deposits associated with Mt. Hood) and the older CRBG aquifers. Newcomb (1969) was the first to document the defining role of the Rocky Prairie thrust fault (then referred to as the Rocky Prairie anticline) as a significant hydraulic barrier to groundwater flow in the study area. Although the current study area is centrally located within the much larger Hood Basin groundwater resources area (Grady, 1983), efforts to understand the effect of the complex geologic on the hydrogeology near Mosier were limited. A detailed study (Lite and Grondin, 1988) of the area immediately to the south of the Rocky Prairie thrust fault (fig. 2) identifies the principal aquifers and their geometry over much of the current study area. This description of hydrogeologic units has been used for this and all subsequent studies (Keinle, 1995; Jervey, 1996).

The geologic map (fig. 2) is a compilation of work by Newcomb (1969), Swanson and others (1981), Bela (1982), Lite and Grondin (1988), Kienle (1995), and Jervey (1996). The primary sources for the refinement of the regional geologic maps were surficial and structural geologic interpretations by Lite and Grondin (1988), Kienle (1995), and Jervey (1996).

Objectives and Approach

To support the evaluation of causes for water level declines in the Mosier area, the study had three objectives:

1. Develop a better understanding of the hydrogeologic framework (the three-dimensional geometry and distribution of hydraulic properties of the aquifer system) (see section Hydrogeologic Framework and appendix A);

2. Estimate major groundwater system water fluxes for use in developing a groundwater system budget (see sections Conceptual Model of the Flow System, Recharge, and Discharge; and appendixes B, C, and D); and

3. Integrate the understanding of the hydrogeologic framework and the water budget into a quantitative tool that can be used to evaluate the causes of water level declines and forecast the effects of management options (see Groundwater-Flow Simulation and appendix E).

Because the geometry of the geologic units controls the storage and movement of groundwater, the first objective was achieved through the development of a three-dimensional geologic model and interpretation of hydrologic data in the context of this model. The second objective was achieved by using a watershed process model (precipitation-runoff model) to estimate the spatial and temporal distribution of recharge, and by conducting a 2-year intensive data collection period during which measurements were made of streamflow, pumping, and vertical borehole leakage in commingling wells. The final objective was achieved by combining the geologic model, a conceptual understanding of flow-controlling features, and the water budget to develop a groundwater-flow simulation model. Historical groundwater-level measurements were augmented with 2 years of intensive measurement to aid in development of the conceptual model of the groundwater-flow system and to provide additional calibration data for the groundwater-flow simulation model.

Hydrogeologic Framework

Groundwater occurs in sediments and rock beneath the land surface. Geologic materials that transmit significant amounts of water are called aquifers, and materials that transmit water poorly are called aquitards. Geologic units generally are delineated based on how and when they were deposited, but a geologic unit may contain both aquifers and aquitards. A saturated aquitard that is areally extensive and serves to confine an adjacent artesian aquifer or aquifers is called a confining unit. Leaky confining units may transmit appreciable water to and from adjacent aquifers. A hydrogeologic framework is constructed by representing the distribution of geologic units and separating, or combining, these units into hydrogeologic units that have similar hydraulic properties.

Although the Mosier area geologic units are well defined at the land surface, their location in the subsurface where groundwater occurs is poorly understood. Using data collected as part of this study and previous studies, the depth, thickness, and extent of important sediment and rock units beneath the study area were mapped. Insufficient subsurface data exists to define the geometry of geologic units to the south of the Chenoweth thrust fault (fig. 2) reliably, so the constructed geologic model does not cover the entire Mosier-Rock-Rowena Creek watershed.

Geologic Setting

The Mosier basin was inundated with flood basalts in Miocene time, followed by deposition of volcaniclastic deposits of mostly Tertiary age (Newcomb, 1969; Swanson and others, 1981; Bela, 1982; and Lite and Grondin, 1988). A small part of the study area is covered with Quaternary fluvial sediments consisting of catastrophic Missoula Floods deposits and modern river and stream deposits. The pre-Miocene basement rock has not been encountered in wells, nor is it exposed in outcrop. Tectonic forces have deformed the system, resulting in faulted and folded basalt.

The geometry of the system is dominated by the Mosier syncline and Columbia Hills anticline (fig. 2) which deform all hard rock units and form the troughs within which the sedimentary overburden is emplaced. The axes of these two features are approximately parallel, with a southwest to northeast trend. As these folds developed, a series of hydraulically important faults also developed, including the Rocky Prairie and Chenoweth thrust faults and a wrench fault whose trace crosscuts the entire watershed starting in the northwest corner of the model area and trending to the southeast (fig. 2). Each of these faults has significant offset over at least a part of their lengths. The wrench fault is associated with the Maupin Trend (Anderson, 1987), and is hereafter referred to as the Maupin wrench fault in this report.

The oldest CRBG lavas were sheet flows that resulted from a high-volume of lava flowing over flatter terrain, which resulted in a laterally extensive continuous coverage under the entire groundwater model area. As the rock was deformed and the syncline-anticline pair developed, the Mosier syncline became a trough through which later CRBG lavas flowed. The geometry of the valleys during deposition and the low volume of lava resulted in flows, called intracanyon flows, which do not cover the entire watershed area. Between periods of deposition of CRBG lavas, sedimentary deposits accumulated on the surface of the previous lava flow, and where these deposits are preserved and covered by a later lava flow, they are called sedimentary interbeds or interbeds. After CRBG volcanism stopped depositing lava in the area, volcaniclastic deposits associated with Cascadian volcanism flowed from the southwest across much of the watershed. The volcaniclastic deposits are highly heterogeneous and poorly delineated, but generally consist of debris flows and volcanic ash.

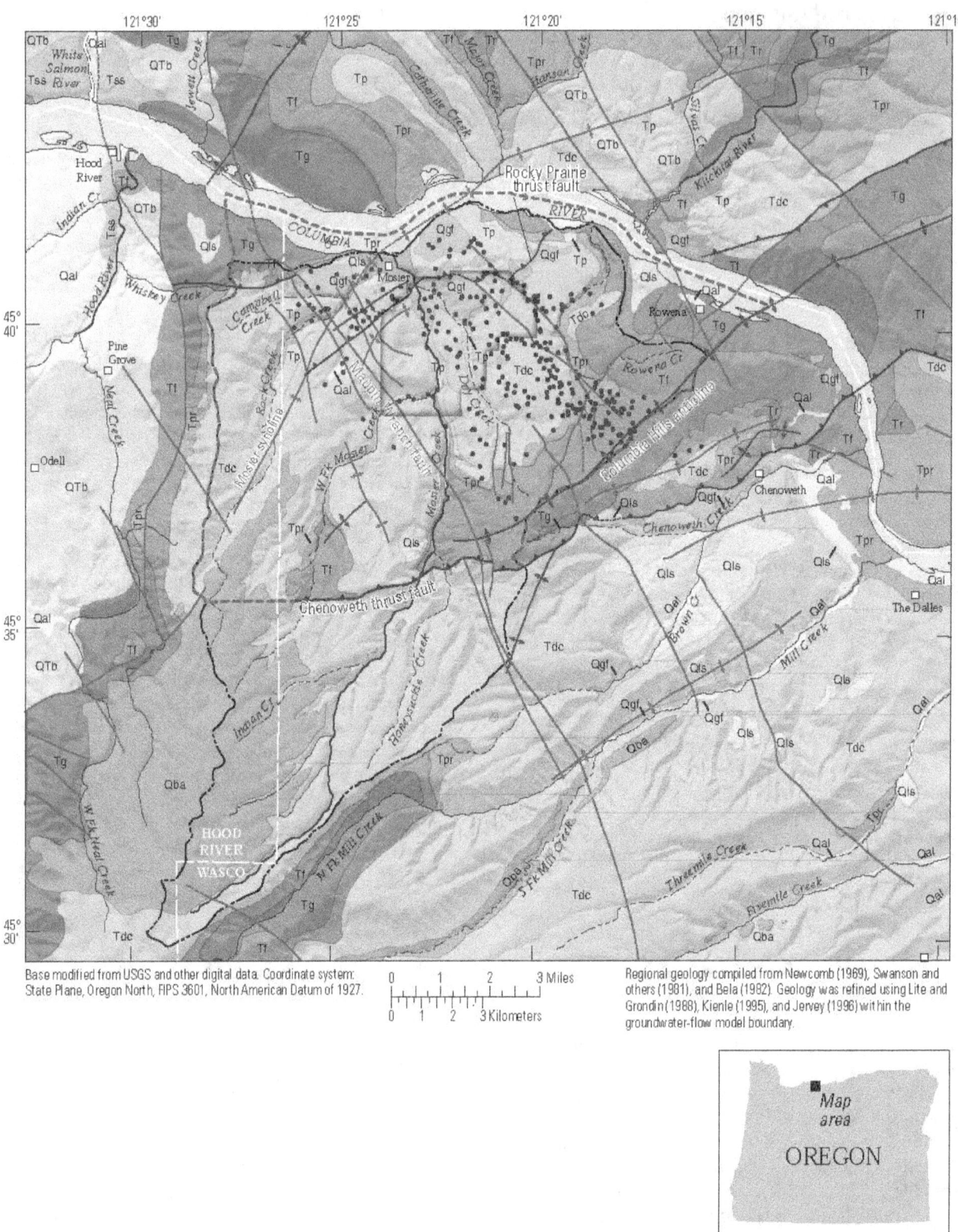

Figure 2. Geology, structural features, and locations of wells used to construct the three-dimensional hydrogeologic framework model of the aquifer system underlying the Mosier, Oregon, study area. The extents of the groundwater-flow model and the watershed precipitation-recharge model also are shown. GIS data are available at http://pubs.usgs.gov/sir/2012/5002/.

EXPLANATION

Surficial geology

Qal — Alluvium consisting of gravel, sand, silt, and clay. Can include some glaciofluvial deposits.

Qls — Landslide deposits resulting from rock slides and block slumps. Also includes small colluvium deposits.

Qgf — Glaciofluvial deposits of coarse, unsorted gravel, sand, and silt from the Pleistocene Missoula Floods.

Qba — Andesite and basalt of Quaternary age, primarily from Mt. Hood.

QTb — Olivine basalt and andesite in undifferentiated Quaternary-Tertiary volcanic units.

Tdc — Chenoweth Formation (of the Dalles Group) includes volcaniclastic and sedimentary rock with laharic deposits, tuff breccia, and fluvial deposits.

Tp — Pomona member of the Columbia River Basalt Group (CRBG). Gray to black, fine grained, porphyritic basalt.

Tpr — Priest Rapids member of the CRBG. Dark gray to black, fine to coarse grained basalt.

Tr — Roza member of the CRBG. Dark gray to black, medium grained, porphyritic basalt.

Tf — Frenchman Springs member of the CRBG. Dark gray to black, fine to medium grained, aphyric to porphyritic basalt.

Tg — Grande Ronde Formation of CRBG. Aphyric, fine grained basalt.

Tss — Conglomerate of Snipes Mountain. Weakly consolidated deposit of fluvial gravel and sand. Formed by ancestral Columbia River.

Anticline — Anticline

Anticline concealed

Fault — Strike-slip fault

Strike-slip fault concealed

Syncline — Syncline

Syncline concealed

Fault — Thrust fault

Thrust fault concealed

— Geologic model and groundwater-flow model boundary

— Mosier-Rock–Rowena Creek watershed boundary

— Perennial stream

— Ephemeral stream

Oregon Water Resources Department (OWRD) Groundwater Administrative Area

• Stratigraphic control well

Figure 2.—Continued

Large floods associated with failure of ice dams near Missoula, Montana, during the last ice age deposited coarse-grained glaciofluvial deposits in a limited area of the lower watershed (fig. 2). The youngest sedimentary deposits in the system are associated with modern erosional processes, and they typically occur near the creeks.

Geologic Model Units

Geologic model units for this study (fig. 3) consist of sedimentary deposits and basalt units of the CRBG, overlying volcaniclastic deposits, and catastrophic flood deposits. The geologic model units used to create a three-dimensional geologic model were selected based on data availability. Mapped or previously identified geologic units were sometimes simplified into simpler geologic model units if data density was insufficient to define the geometry of the units.

The geologic deposits overlying the CRBG aquifers consist of a variety of alluvial and volcaniclastic deposits, referred to hereafter as overburden. These deposits were grouped into younger Glaciofluvial Deposits and (mostly) older Undifferentiated Overburden (fig. 3) based on preliminary groundwater flow modeling results. The thickness of the overburden is highly variable. No estimates of maximum thickness are available because the thickest sequences are likely in the trough of the Mosier syncline, and no thickness data are available near the syncline axis.

Below the overburden, a series of CRBG lava flows covers the watershed. Because water wells typically penetrate the minimum depth in the aquifer system that meets water-usage needs, more information is available for the shallower units at any given location. The three youngest lava units (Pomona, Lolo, and Rosalia) and the two uppermost interbeds (Selah and Quincy-Squaw Creek) are identified with regularity and reasonable confidence in most well logs, allowing identification of each of these units in the geologic model. Each of these lava units consists of a single intracanyon flow that partially covers the study area. Flow thicknesses are variable, pinching out at the margins, but with typical thickness of about100 ft in many areas.

The composition of the sedimentary interbeds is highly variable. The Selah interbed lies between the Pomona and Lolo Basalt units and the Quincy-Squaw Creek interbed lies between the Rosalia and Sentinal Gap Basalt units. Sedimentary interbed thickness is highly variable and may be discontinuous over short distances, with thickness depending on paleotopography of the surface over which the overlying basalt flowed. The Selah interbed thickness is apparently correlated (with high variability) to thick sections of the Pomona Basalt, which is likely because sedimentary deposits tend to be thickest in valley bottoms that the lava filled. Recorded thicknesses of the Selah interbed in well logs range from 0 to about 100 ft. The Quincy-Squaw Creek interbed thickness typically ranges between 10 and 30 ft in well logs, with no apparent correlation to overlying lava unit thickness.

The next youngest basalt unit is the Roza, a single flow deposited during the same period when the Quincy-Squaw Creek interbed was deposited (Tolan and others, 2009). Some of the Quincy-Squaw Creek interbed deposits may be older and some may be younger than the Roza flow. For modeling purposes, the Roza basalt unit was assumed to underlie the interbed (fig. 3). The Roza flow is of limited areal extent, occurring only near Rowena Creek (Lite and Grondin, 1988).

Three of the Frenchman Springs units are mapped in the Mosier area (Tolan and others, 2009), accounting for at least four lava flows: one Sentinal Gap flow, two Sand Hollow flows, and one or more Gingko flows (Kenneth Lite, Oregon Water Resources Department, written commun., 2010). Because of insufficient data on the geometry of the Roza flow and the Frenchman Springs flows, these flows have been lumped into the Frenchman Springs geologic model unit (fig. 3). Almost all wells that penetrate the Frenchman Springs units are near the crest of the Columbia Hills anticline. In this area, where the Roza likely is absent, the total thickness is estimated at about 400 ft.

The Grande Ronde Basalt likely underlies the entire study area, even though it is only identified in wells near the Columbia Hills anticline. The total number of flows and total thickness are not known, although a thick sequence of Grande Ronde Basalt is exposed in the Columbia River Gorge on the northeast boundary of the study area. The top of this unit forms the lower bound for the geologic model.

Three-Dimensional Geologic Model

The hydrogeologic framework was developed using a three-dimensional geologic model (figs. 4 and 5). The geologic model was constructed for the area where geologic maps and geologic interpretation of 318 well logs from previous studies (Newcomb, 1969; Grady, 1983; Kienle, 1995; Jervey, 1996) provided sufficient information to define the three-dimensional geometry of the geologic units constituting the aquifer system (fig. 2). To the south of the Chenoweth thrust fault, because volcaniclastic deposits cover the area, the geometry of the underlying geologic units is poorly understood. As a result, the geologic model and the derivative groundwater-flow simulation model domains do not extend to the south of the Chenoweth thrust fault.

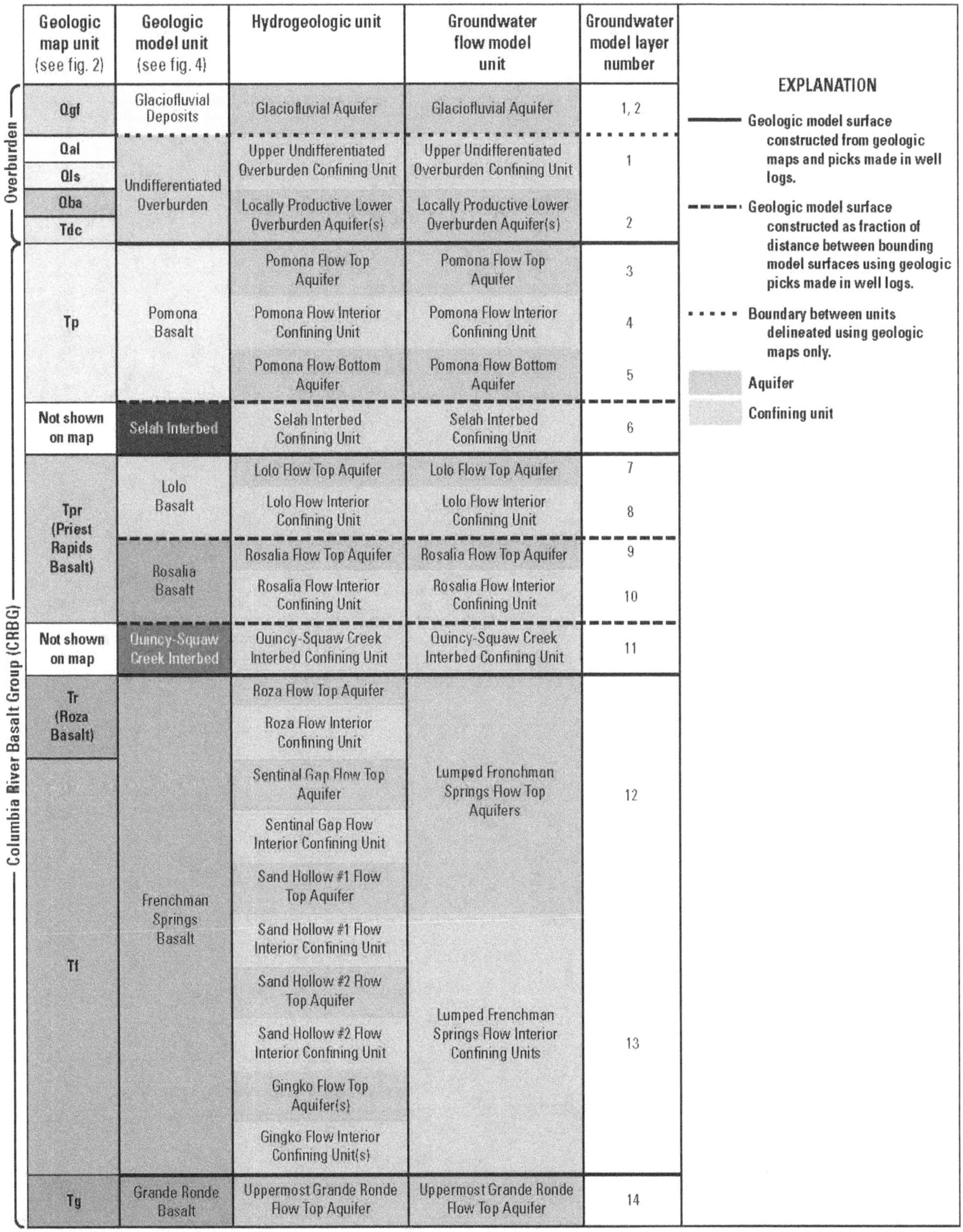

Geologic map unit (see fig. 2)	Geologic model unit (see fig. 4)	Hydrogeologic unit	Groundwater flow model unit	Groundwater model layer number	
Overburden Qgf	Glaciofluvial Deposits	Glaciofluvial Aquifer	Glaciofluvial Aquifer	1, 2	**EXPLANATION** —— Geologic model surface constructed from geologic maps and picks made in well logs.
Qal Qls Qba Tdc	Undifferentiated Overburden	Upper Undifferentiated Overburden Confining Unit	Upper Undifferentiated Overburden Confining Unit	1	‐ ‐ ‐ Geologic model surface constructed as fraction of distance between bounding model surfaces using geologic picks made in well logs.
		Locally Productive Lower Overburden Aquifer(s)	Locally Productive Lower Overburden Aquifer(s)	2	· · · · Boundary between units delineated using geologic maps only.
Columbia River Basalt Group (CRBG) Tp	Pomona Basalt	Pomona Flow Top Aquifer	Pomona Flow Top Aquifer	3	▨ Aquifer
		Pomona Flow Interior Confining Unit	Pomona Flow Interior Confining Unit	4	▨ Confining unit
		Pomona Flow Bottom Aquifer	Pomona Flow Bottom Aquifer	5	
Not shown on map	Selah Interbed	Selah Interbed Confining Unit	Selah Interbed Confining Unit	6	
Tpr (Priest Rapids Basalt)	Lolo Basalt	Lolo Flow Top Aquifer	Lolo Flow Top Aquifer	7	
		Lolo Flow Interior Confining Unit	Lolo Flow Interior Confining Unit	8	
	Rosalia Basalt	Rosalia Flow Top Aquifer	Rosalia Flow Top Aquifer	9	
		Rosalia Flow Interior Confining Unit	Rosalia Flow Interior Confining Unit	10	
Not shown on map	Quincy-Squaw Creek Interbed	Quincy-Squaw Creek Interbed Confining Unit	Quincy-Squaw Creek Interbed Confining Unit	11	
Tr (Roza Basalt)	Frenchman Springs Basalt	Roza Flow Top Aquifer	Lumped Frenchman Springs Flow Top Aquifers	12	
		Roza Flow Interior Confining Unit			
		Sentinal Gap Flow Top Aquifer			
		Sentinal Gap Flow Interior Confining Unit			
Tf		Sand Hollow #1 Flow Top Aquifer			
		Sand Hollow #1 Flow Interior Confining Unit	Lumped Frenchman Springs Flow Interior Confining Units	13	
		Sand Hollow #2 Flow Top Aquifer			
		Sand Hollow #2 Flow Interior Confining Unit			
		Gingko Flow Top Aquifer(s)			
		Gingko Flow Interior Confining Unit(s)			
Tg	Grande Ronde Basalt	Uppermost Grande Ronde Flow Top Aquifer	Uppermost Grande Ronde Flow Top Aquifer	14	

Figure 3. Relation of geologic units to hydrogeologic units and groundwater-flow model units in the Mosier, Oregon, study area.

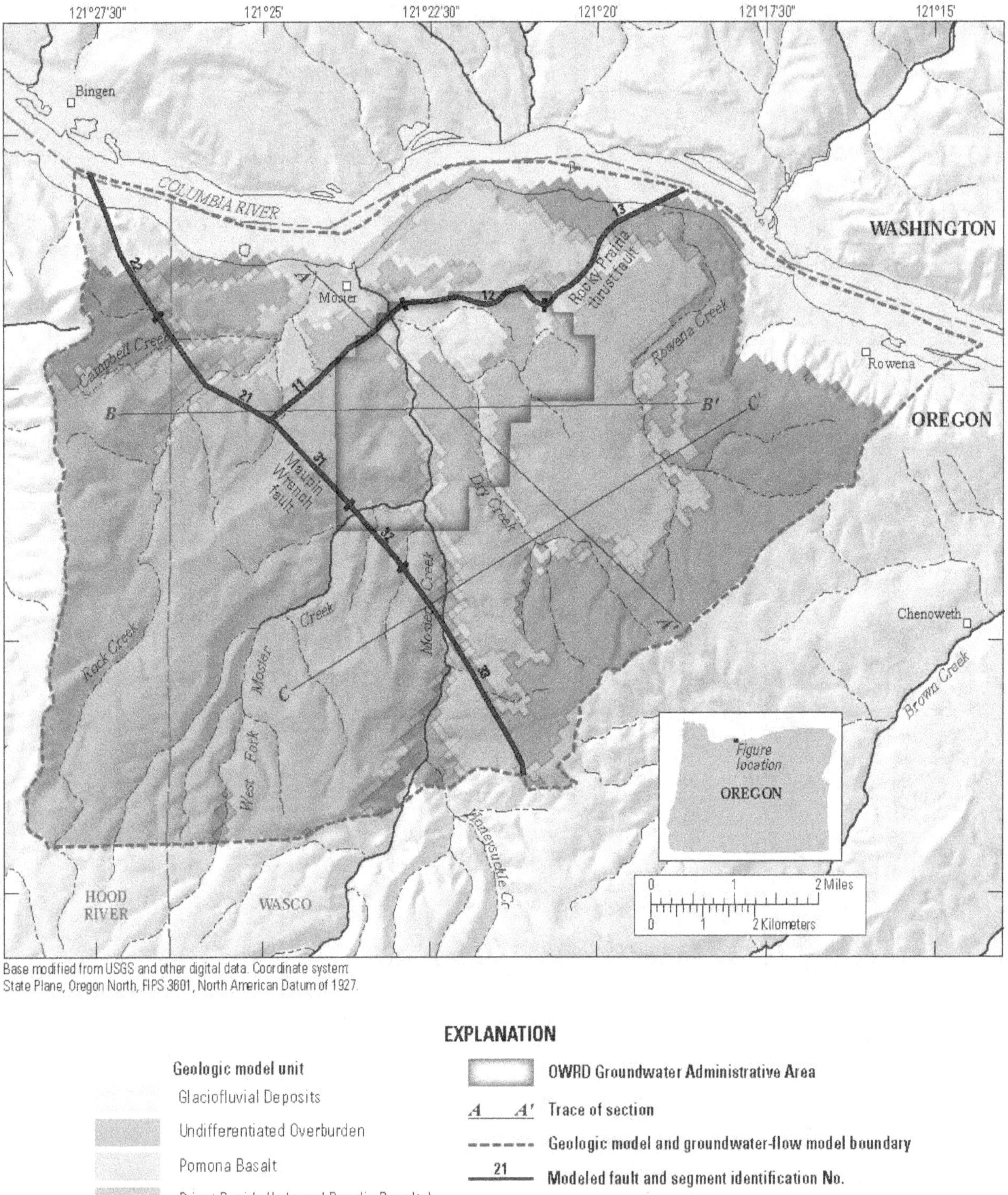

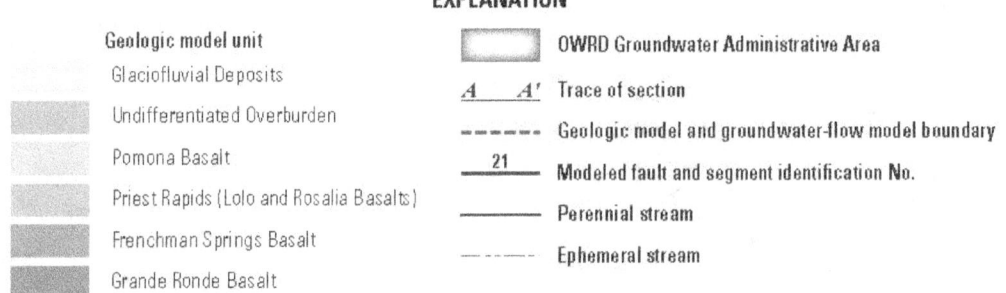

Base modified from USGS and other digital data. Coordinate system
State Plane, Oregon North, FIPS 3601, North American Datum of 1927.

EXPLANATION

Geologic model unit

Glaciofluvial Deposits

Undifferentiated Overburden

Pomona Basalt

Priest Rapids (Lolo and Rosalia Basalts)

Frenchman Springs Basalt

Grande Ronde Basalt

OWRD Groundwater Administrative Area

A   *A'*   Trace of section

- - - - - -   Geologic model and groundwater-flow model boundary

21   Modeled fault and segment identification No.

Perennial stream

- - - - -   Ephemeral stream

Figure 4. Surficial expression of geologic model units as represented in the 500-foot flow model grid for the Mosier, Oregon, study area.

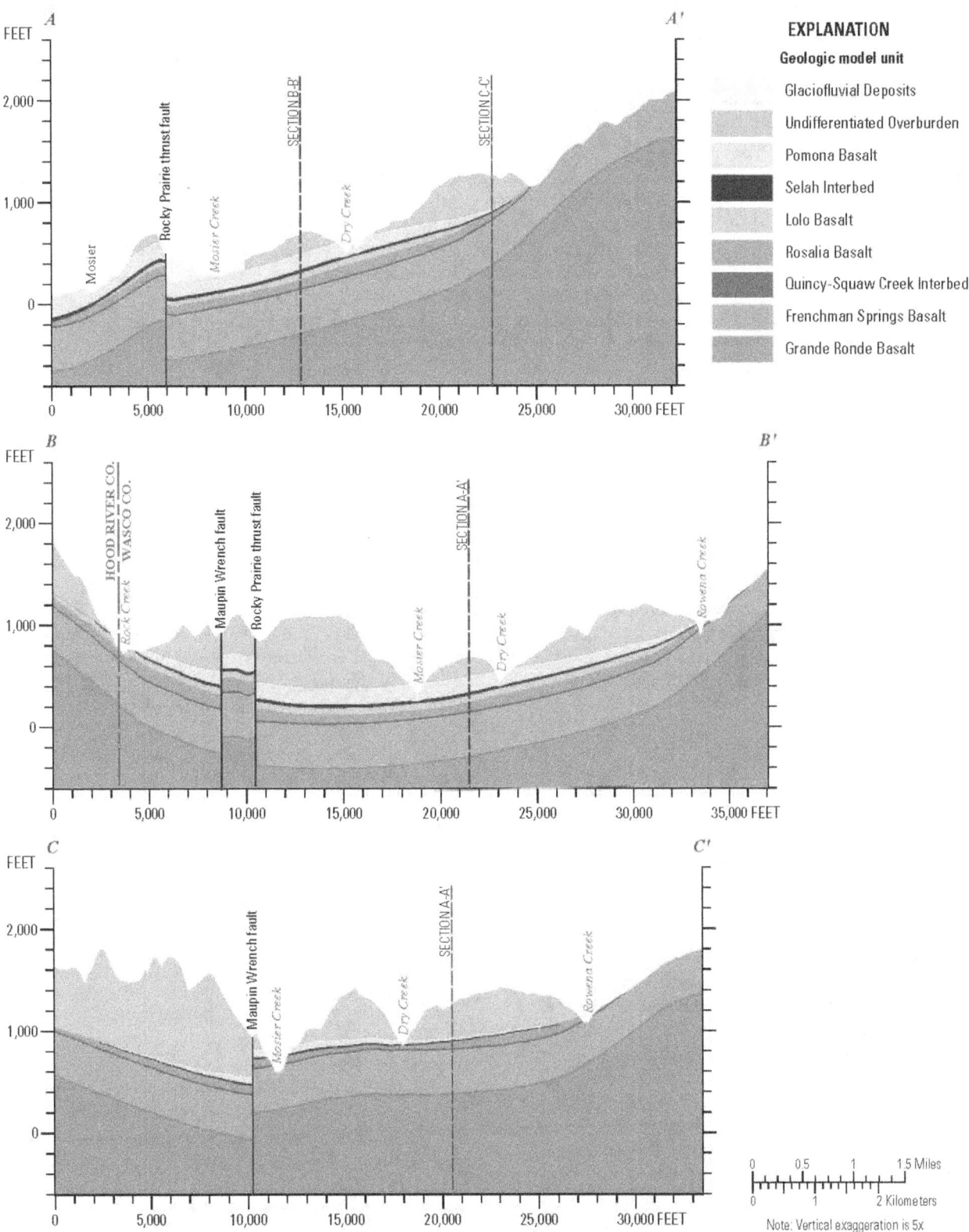

Figure 5. Cross sections through the geologic model for the Mosier, Oregon, study area.

Because potential errors exist in all of the data, trend interpolation methods were used to develop the three-dimensional geologic model from the data. Inductive methods were used for construction of the geologic map and interpretation of well stratigraphy, where geologists identified the likely location of geologic contacts based on contacts identified at other locations. For some wells, the geologic interpretation of stratigraphy (termed 'geologic pick') differed between studies, so one or both conflicting interpretations of the geology contain errors. For wells with conflicting geologic picks, a single "best" pick was made using available data. Geologic unit tops and bottoms were simulated using two-dimensional surface trend models to ensure that the final geologic model matches most of the data well, preserving the important features of the system that control the storage and transmission of groundwater. The details of constructing the surfaces and the geologic model are described in appendix A.

Preliminary groundwater-flow simulation results indicated that the hydraulic conductivity of the glaciofluvial deposits was possibly an important parameter for understanding aquifer leakage through commingling wells, so zonation was used to separate the glaciofluvial deposits from the remainder of the overburden. Wherever glaciofluvial deposits exist, the geometry of any older buried overburden is poorly understood, so it was assumed that if glaciofluvial deposits are mapped at land surface, they are the only overburden unit present (fig. 5A–A'). This is a poor geologic assumption, but it allows testing of the role of the glaciofluvial deposits in the groundwater-flow simulation model.

Hydrogeologic Units

Following creation of the three-dimensional geologic model, geologic units were divided into hydrogeologic units, where the flow controlling features were identified (fig. 3). This conceptual model allowed the identification of geologic features that are believed to control the response of the groundwater system.

Hydrogeologic units were defined based on their hydraulic characteristics. If adjacent geologic units have similar abilities to store and transmit water, then they can be grouped into a single hydrogeologic unit. Conversely, if a geologic unit has zones of significantly different hydraulic character, then geologic units can be divided into multiple hydrogeologic units. Geologic material that is very permeable to water is called an aquifer, and significantly less permeable units are called aquitards. Laterally extensive aquitards are called confining units. High permeability corresponds to high hydraulic conductivity, a measure of how easily water is transmitted through geologic material. Similarly, low permeability corresponds to low hydraulic conductivity. For this study, 23 hydrogeologic units (aquifers and confining units) were defined (fig. 3).

The overburden geologic units were divided into three hydrogeologic units based on the hydrologic properties of these units and their potential influence on important groundwater flow processes (fig. 3): the glaciofluvial aquifer, the upper undifferentiated overburden-confining unit, and the locally productive lower overburden aquifer. The uppermost hydrogeologic unit is the glaciofluvial aquifer, consisting of very permeable gravel and other coarse sediments deposited during the Missoula Floods. These deposits are of limited extent (fig. 2), and where they occur to the south of the Rocky Prairie thrust fault, the permeability may control the rate at which water leaking vertically through commingling wells would return to Mosier Creek. This unit was separated from the undifferentiated overburden in the geologic model to evaluate the role of the glaciofluvial deposits in restricting the flow from commingling wells.

The remainder of the overburden is undifferentiated, but the largest part of this geologic model unit consists of Cascadian volcaniclastic deposits that are older than the glaciofluvial deposits. Previous investigators (Newcomb, 1969; Lite and Grondin, 1988) recognized that, although these deposits typically have low permeability, coarser deposits forming productive aquifers may occur in the lower parts of the unit. For this reason, the undifferentiated overburden geologic model unit is divided conceptually into an upper confining unit and a lower aquifer that may be discontinuous.

Generally, each CRBG lava flow consists of a dense flow interior and irregular flow tops and flow bottoms with a variety of textures (fig. 6) (Reidel and others, 2002). Flow top textures are formed as the lava develops a crust while the liquid center continues to flow. Flow bottom textures are controlled by the lava properties (for example, temperature and chemical composition) and the properties of the surface over which the lava is flowing. A variety of joint patterns, fractures, and lithologic textures can occur in any single basalt flow. Although flow interiors have joints and fractures, they typically do not transmit water easily. Flow tops and bottoms are commonly vesicular or brecciated, and they may or may not be permeable. Local permeability of flow tops and bottoms may be highly variable over short distances as a result of depositional processes, but the complex connectivity of the open conduits tends to be high over long distances, resulting in highly transmissive aquifers at the regional scale. The variability in lithologic textures implies that even though a flow top or bottom is intersected when drilling, there is no guarantee that this zone will be open and connected to the aquifer system. Within the study area, flow tops generally tend to transmit water easily, forming productive aquifers, but the only documented transmissive flow bottom is at the base of the Pomona Basalt flow (Lite and Grondin, 1988). The Pomona Flow Bottom aquifer does not occur at all locations where the Pomona Basalt occurs, but to the south of the Rocky Prairie thrust fault, it was estimated to cover an area of 4–6 mi^2, generally coincident with the OWRD groundwater administrative area (fig. 1). In this area, the aquifer may

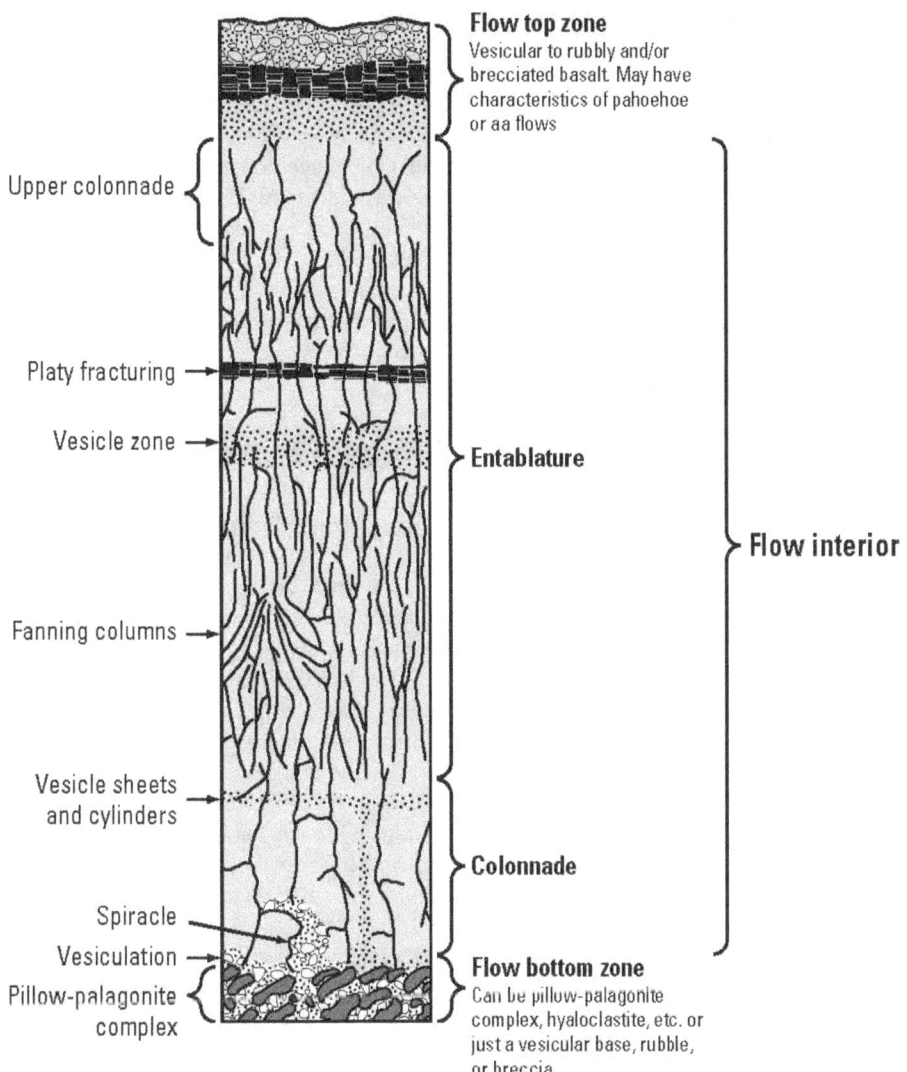

Flow top zone
Vesicular to rubbly and/or
brecciated basalt. May have
characteristics of pahoehoe
or aa flows

Upper colonnade

Platy fracturing

Vesicle zone

Entablature

Flow interior

Fanning columns

Vesicle sheets
and cylinders

Colonnade

Spiracle

Vesiculation

Flow bottom zone
Can be pillow-palagonite
complex, hyaloclastite, etc. or
just a vesicular base, rubble,
or breccia.

Pillow-palagonite
complex

Figure 6. Geologic features that control flow and storage in Columbia River Basalts (from Reidel and others, 2002).

be as much as 40 ft thick (Lite and Grondin, 1988). The transmissive flow bottom is postulated to have formed when the lava flowed over wet sediments in the paleo-valley bottom.

As a group, the CRBG is a stack of laterally extensive lava flows with relatively thin permeable, productive zones at flow tops and flow bottoms separated by relatively thick flow interiors of low permeability. Individual CRBG lava flows may be tens to hundreds of feet thick, with a typical thickness of about 100 ft in the study area. Thickness of each part of each flow is highly variable locally, but the thin permeable aquifers commonly occupy about 10 percent of the total thickness. The aquifer system can transmit and yield water easily from the thin flow tops and bottoms, but has low storage capacity in the flow interiors, which make up a large part of the aquifer system. Flow interiors have low permeability and low storage characteristics, and they form effective confining units between permeable flow tops.

Sedimentary interbeds between CRBG lava flows are porous and able to store water but are less permeable than the adjacent basalt aquifers, so they form confining units in the Mosier study area. The combined thickness of flow top, interbed, and an overlying flow bottom is called an interflow. If a continuous interbed exists between a permeable flow bottom and permeable flow top, the interbed typically functions as a confining unit, dividing the interflow into two aquifers. In the absence of an interbed, the flow top and overlying flow bottom are hydraulically indistinguishable, so a single aquifer exists. Whether the single aquifer is comprised of only a permeable flow top or the combination of a permeable flow top and permeable flow bottom, the hydrogeologic nature of the aquifer is the same, and these aquifers are designated as flow top aquifers in the terminology of this report (fig. 3).

Three-Dimensional Hydrogeologic Framework

A three-dimensional hydrogeologic framework suitable for input to a groundwater-flow model was constructed by dividing the geologic model into groundwater-flow model units. To the maximum extent practicable, each geologic model unit was divided into separate groundwater-flow model units representing hydrogeologic units (fig. 3). This was accomplished for all geologic model units except the Frenchman Springs unit, which was grouped into groundwater-flow model units representing the bulk properties of the five or more basalt flows within the unit. Each groundwater-flow model unit is represented as a single layer in the groundwater-flow model with the exception of the overburden.

The overburden was divided into two layers in two zones in the groundwater-flow model. The undifferentiated overburden was divided into an upper confining unit and a lower aquifer in one zone and the homogeneous glaciofluvial aquifer occupied both layers in the other zone. The geometry of the aquifer that locally occurs at the base of the Chenoweth Formation is poorly understood, but generally is assumed to be thin relative to the entire thickness. To allow this unit to be represented in the groundwater-flow model, it was arbitrarily assumed that the lower undifferentiated overburden aquifer occupied 10 percent of the total thickness. The glaciofluvial aquifer layers share the same percentage split in thickness as the undifferentiated overburden, but both layers are assigned the same properties, so that this unit is modeled as homogeneous.

Each major sedimentary interbed is represented as a single hydrogeologic unit, and the Pomona, Lolo, and Rosalia Basalt flows were subdivided into aquifers and confining units as described in the Hydrogeologic Units section. The basalt flow top aquifers were assumed to occupy 10 percent of the total thickness. The Pomona Basalt flow bottom aquifer geometry was modeled to match estimates of extent and thickness estimated by Lite and Grondin (1988). The areal extent was defined by identifying a thickness threshold such that the thickness of Pomona basalt exceeding this threshold occupies about 4 mi^2 to the south of the Rocky Prairie thrust fault near the OWRD groundwater administrative area. Thickness of this aquifer was defined as a fraction of the total thickness exceeding the threshold such that the thickest part of the aquifer in the OWRD administrative area is approximately 20 ft thick. The remainder of the thickness of each basalt unit was defined as a flow interior confining unit.

Conversely, the Frenchman Springs geologic model unit represents a sequence of five or more basalt flows. For flow modeling purposes, this unit is divided into two flow model units, with each unit represented by a single flow model layer. The upper unit represents the group of flow top aquifers that are associated with the Frenchman Springs geologic model layer. The lower unit represents the group of low permeability flow interiors. The lumped flow top aquifer is modeled as the upper 10 percent of the total thickness, with the lower 90 percent being modeled as a lumped confining unit.

The Grande Ronde Basalt unit top was the lower bound of the geologic model. A single 20 ft thick groundwater model layer was used to simulate a single flow top aquifer associated with the Grande Ronde aquifer system. The flow interior below this aquifer is not simulated in the groundwater-flow model because it is likely a barrier to flow and no wells penetrate it.

Supporting data and additional details of division of the geologic model into groundwater-flow model units are contained in appendix A.

Groundwater-Flow System

Conceptual Model of the Flow System

The study area and conceptual model of Lite and Grondin (1988) was extended to likely natural hydrologic boundaries for the purposes of groundwater-flow simulation. The major hydrogeologic processes and concepts used to define the study area are described in this section. Data collection, groundwater recharge, movement, discharge, and water-level changes are summarized in the following sections.

Lite and Grondin (1988) presented a conceptual model of flow in the study area along the transect A–A' (fig. 5). The principal hydrogeologic features considered were the basalt aquifers and their interactions with the Rocky Prairie thrust fault and incised creeks, primarily Mosier Creek. Because the Mosier Creek gradient is less than the dip of the CRBG units, the creek cuts across several basalt aquifers along its length, with lower aquifers exposed at higher elevations in the watershed (compare B–B' with C–C', fig. 5). The Rocky Prairie thrust fault acts as a groundwater-flow barrier, causing groundwater recharge in the uplands to fill the basalt aquifers until springs and seeps form where aquifers intersect the land surface (fig. 7).

For the current study, the area covered by the Lite and Grondin (1988) conceptual model was extended to include natural hydrogeologic boundaries appropriate for groundwater flow simulation. The southeastern boundary is coincident with the Columbia Hills anticline where a combination of the anticline and the draped Chenoweth thrust fault suggests that water recharged to aquifers will flow away from the anticline on both sides, implying that groundwater and surface water do not flow laterally across the anticline.

The eastern extent of the Lite and Grondin conceptual model was Rowena Creek and an associated mapped fault. Because the hydrogeologic role of Rowena creek is not readily apparent, and because it may have a hydrogeologic effect similar to Mosier Creek, the study area boundary was moved further east to the Columbia River Gorge and the Columbia Hills anticline, encapsulating the entire Rowena Creek drainage.

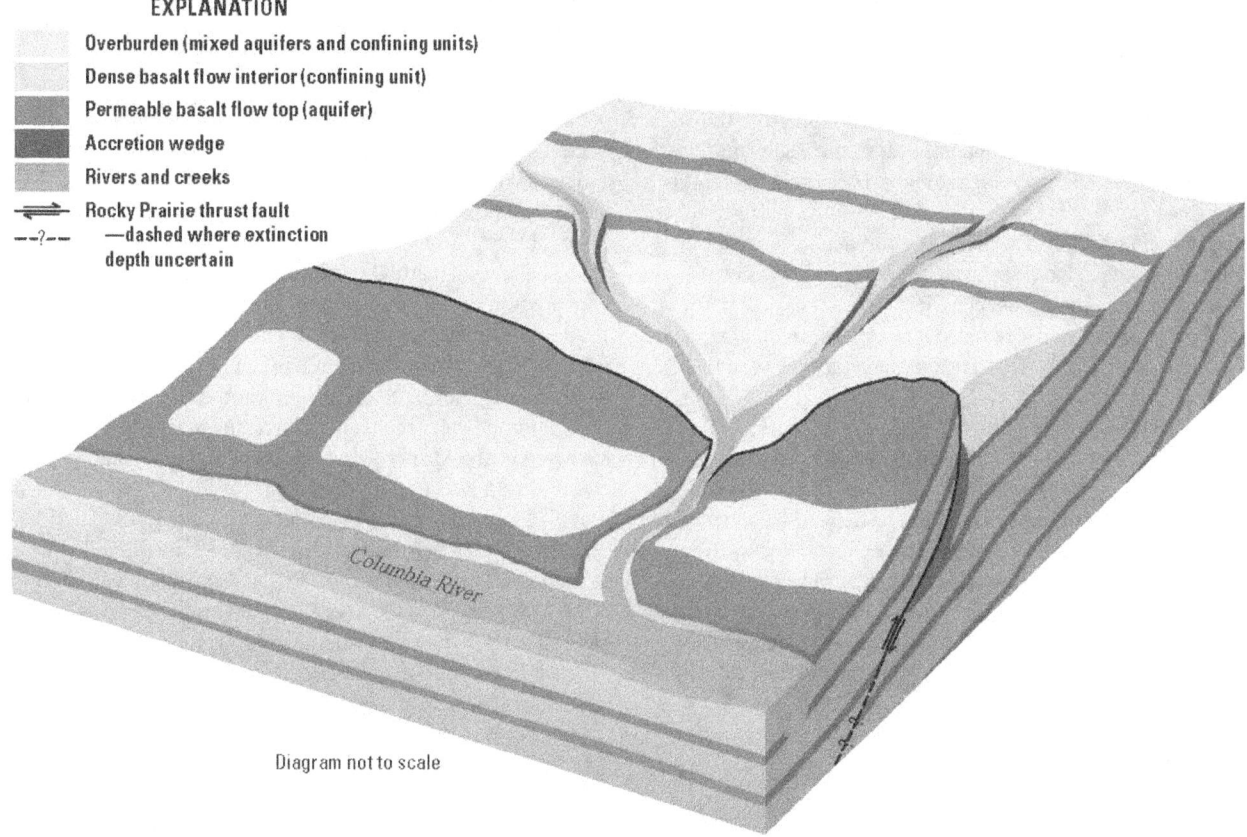

EXPLANATION

Overburden (mixed aquifers and confining units)

Dense basalt flow interior (confining unit)

Permeable basalt flow top (aquifer)

Accretion wedge

Rivers and creeks

Rocky Prairie thrust fault

—?— —dashed where extinction depth uncertain

Columbia River

Diagram not to scale

Figure 7. The conceptual model of groundwater flow in the Mosier, Oregon, study area.

The western extent of Lite and Grondin was a north-south line generally corresponding to the westernmost edge of the Rocky Prairie thrust fault and the extent of the available data. Rock Creek likely has a similar hydrogeologic effect on parts of the groundwater system as Mosier Creek, so the remainder of the study area was defined by adding the combined total drainage area of Mosier and Rock Creeks (fig. 1). This new boundary is coincident with the ridge to the west of Rock Creek, which is paralleled by a high offset normal fault. The combination of ridge and fault is a likely barrier to groundwater flow, making this boundary a barrier to groundwater and surface water.

The Columbia River forms the entire northern boundary of the study area. All groundwater and surface water in the study area naturally drain from the uplands toward the river, and the river crosscuts all of the basalt aquifers of interest along some part of its length.

In this extended study area, two additional faults were identified as potential barriers to flow based on geologic modeling: the Maupin wrench fault and the Chenoweth thrust fault. Mosier and Rock Creeks intersect multiple aquifers and flow across these potential flow barriers, creating a complex geometry between aquifers and creeks.

The part of the study area to the south of the Chenoweth thrust fault (fig. 2) is completely covered with overburden. The geometry of the aquifer system is unknown and no groundwater-level data exist. It was assumed that the overburden is similar to the low permeability units described in the northern areas of the study area and that the Chenoweth thrust fault is likely a hydrogeologic barrier similar to the Rocky Prairie thrust fault. Given this combination of overburden and thrust fault, much of the groundwater above the Chenoweth thrust fault likely drains into the creeks above the fault.

Rainfall, and consequently recharge, varies gradationally across the watershed, with more rainfall occurring, and presumably more recharge entering, the aquifer system in upland areas. The older sheet flow basalts that underlie the entire study area are only exposed at land surface in uplands near the structural anticline, allowing recharge into these deeper units (fig. 2). Downslope, parts of the younger intracanyon CRBG lava flows are exposed at land surface, and parts are buried beneath the overburden, which allows recharge into the younger basalt aquifers.

Water enters the CRBG aquifers, flowing from the uplands towards the Columbia River. The Rocky Prairie thrust fault interrupts the lateral continuity of several shallow CRBG aquifers, forming a barrier to flow. Hydraulic heads in the upper CRBG aquifers north of the fault are similar to the stage of the Columbia River, although heads in the same aquifers to the south are hundreds of feet higher. Although the fault is a barrier to flow, it is likely imperfect, so some groundwater may flow through or past the fault while other groundwater drains into creeks from springs and seeps. No groundwater-level data exists for the deeper CRBG aquifers to the north of the Rocky Prairie thrust fault, so it is not known to what extent the thrust fault acts as a barrier to groundwater flow in these deeper aquifers.

At locations where aquifers intersect the creek, groundwater and surface water are in direct connection with each other (fig. 7). If the groundwater level is below the creek, water leaks from the creek into the aquifer, and vice versa. If the aquifer is exposed above the creek level, then water may also flow out of the aquifer through springs and seeps, draining into the creek. All these conditions occur along the length of Mosier Creek between the Chenoweth and Rocky Prairie thrust faults, depending on location and time of year.

Recharge

Three sources contribute recharge to the basalt aquifers. The primary source of recharge is precipitation that infiltrates past the plant root zone to the groundwater system. Second, part of the water pumped for irrigation and domestic usage may return to the groundwater system by infiltration. Third, leakage from streams to the groundwater system can occur in locations where streambeds are permeable and stream water levels are higher than the hydraulic head in the connected aquifer. Recharge infiltration past the root zone from precipitation, irrigation, and domestic usage was estimated, and recharge from streams was estimated during the groundwater simulation process.

Recharge from precipitation was estimated by two independent methods. The primary method, and the method that provided an estimate of recharge over the entire study area, was based on Precipitation Runoff Modeling System (PRMS) (Leavesley and others, 1996). PRMS is a watershed model that balances the input of precipitation with numerous outputs, including evaporation, runoff, and of primary interest for groundwater flow simulations, water that recharges the groundwater system. Calibration of PRMS was accomplished for the part of the Mosier Creek basin upstream of the stream-gaging station (streamflow

measurement site 4 on fig. 1) by adjusting model parameters to minimize the difference between simulated and observed daily streamflows. The PRMS model was then expanded in space and time to include the entire study area and the period of groundwater-flow simulation. The simulated values of streamflow were compared to streamflow measurements and seepage estimates made at 14 additional streamflow measurement sites (fig. 1) to ensure the expanded model is a reasonable representation of the entire study area.

The second method to estimate recharge from precipitation uses a computer program, RORA (Rutledge, 1998), to estimate the part of each peak in the streamflow record contributed by flow through the groundwater system. Applied over a long period, the program estimated the mean rate of groundwater recharge that returns to streamflow upstream of the Mosier Creek gaging station (streamflow measurement site 4 on fig. 1). Details of PRMS and RORA are provided in appendix B.

The spatial distribution of annual average groundwater recharge was estimated using PRMS for the period 1955 to 2007. The average value over the entire study area was 9.6 in. (41,100 acre-ft/yr) with recharge varying from about 4 in. in the eastern part of the study area to about 19 in. in the southern upland area. The annual average groundwater recharge to the drainage area upstream of the Mosier Creek gaging station (fig. 1) was estimated at 9.7 in. using PRMS, compared to 13.6 in. estimated using RORA (appendix B.3). This result suggests a range of possible recharge values for the study area, with PRMS providing a relatively conservative lower estimate of recharge compared to RORA.

The final component of recharge is return flow from pumping for irrigation and domestic usage. Water pumped for use by rural residents on small acreages is either applied to lawns and gardens, or used for household needs. Unless over-watering occurs, most of the water applied to lawns and small gardens is consumptively used by evapotranspiration by plants, whereas household drinking and wash waters are non-consumptively used, returning to the uppermost aquifer through septic drain fields. Because typical rural residential-exempt water wells are shallow, tapping the uppermost aquifer, most septic drain water is assumed to return to the aquifer being pumped, indicating a negligible net change in water in the uppermost aquifer resulting from pumping for non-consumptive uses. For this reason, only consumptive use pumping was estimated for rural residential small acreages for representation in the groundwater simulation model (see Pumping of Groundwater), and recharge and non-consumptive pumping from rural residential wells was not simulated.

Irrigation water applied in excess of plant requirements returns to shallow aquifers by percolating through the root zone past the point where plants access the water. Groundwater recharge from irrigation into the principal CRBG aquifers is assumed to be negligible for two reasons. In the relatively small area supplied by large irrigation wells (generally coincident with the OWRD groundwater administrative area, fig. 1), hydraulic heads in the confined basalt aquifers were significantly higher than the overburden water table aquifers, indicating that the amount of recharge to basalt aquifers from irrigation is negligible. This recharge likely enters the overburden aquifers and returns to nearby streams. Additionally, the amount of recharge from irrigation is less than 1 percent of recharge from precipitation because the estimated 740 acre-ft of irrigated water applied to all crops in 2006 (see section Pumping of Groundwater) was 1.8 percent of the estimated average annual recharge due to precipitation (41,100 acre-ft). The high-efficiency sprinkler systems used in the area resulted in only a small fraction of this water infiltrating into the groundwater system.

Groundwater Flow Direction

Groundwater-Level Monitoring Network

Groundwater levels provide a measure of hydraulic head and water stored in the aquifer system. Hydraulic head is a measure (in units of feet above a datum) of the potential to cause flow due to gravity and water pressure. Groundwater flows from high to low hydraulic head.

A network of wells (table 1, fig. 8) was established by USGS and OWRD to monitor changes in groundwater levels over a 2-year intensive period and for comparison with historical levels. The monitoring network is limited to the eastern side of the study area, in part because this is the area where significant groundwater-level declines have been observed, and because few wells exist in the study area west of the Maupin wrench fault and south of the Chenowith thrust fault (fig. 2). Water levels were collected quarterly (4 wells), bimonthly (26 wells), and continuously (7 wells, measurements recorded bihourly) in a network of 37 wells representative of the aquifers in the study area. These wells were privately owned domestic, irrigation, and unused wells where owner permission was granted and were selected to represent each aquifer over the maximum lateral extent possible.

Many of the groundwater-level measurements in the study area are from wells that are potentially open to multiple aquifers. The groundwater level in each of these wells is a composite hydraulic head, representing a flow-weighted combination of hydraulic heads that occur separately in each of the aquifers.

Hydraulic Gradients and Groundwater Movement

The hydraulic gradients and groundwater movement in the study area are controlled by flow barriers associated with the geology. High offset faults interrupt the lateral continuity of the thin basalt aquifers, forming effective barriers to flow, resulting in high gradients across the faults. Laterally extensive, thick confining units separate the basalt aquifers resulting in high vertical hydraulic gradients.

Pre-1970 water levels in shallow basalt wells south of the Rocky Prairie thrust fault were at a water-level elevation of about 475 ft (figs. 9, and 10), and shallow basalt wells north of the fault had water levels between 70 and 90 ft, similar to the Columbia River stage (about 70 ft). Since 1970, groundwater levels have steadily declined to the south of the Rocky Prairie thrust fault, reducing the gradient across the fault by about 175 ft.

Within each aquifer, the hydraulic head is higher in the uplands than near the Rocky Prairie thrust fault. Horizontal gradients are smaller near the thrust than in the uplands. The reasons for this are not clear, but three potential contributing factors have been identified. First, the transmissivity of younger basalts is known to be high near the OWRD management area in the watershed with possible lower transmissivity as these aquifers extend toward the Columbia Hills anticline (Lite and Grondin, 1988). Second, most water-level measurements near the anticline are from Frenchman Springs aquifers, and lower in the watershed, measurements are more frequently from the younger aquifers, indicating that the higher gradient may be the result of the Frenchman Springs aquifers being less transmissive. Third, recharge is higher in the uplands where the Frenchman Springs geologic unit is exposed, which would also result in a steeper hydraulic gradient. Lite and Grondin (1988) provide hydraulic head maps for the Pomona and Priest Rapids aquifers. The maps are complicated, with attempts to account for composite heads and seasonal changes near streams. However, the general patterns summarized here and implied by the conceptual model hold true.

The highest vertical gradient measured between any two adjacent aquifers south of the Rocky Prairie thrust fault is a head difference of about 70 ft across the Selah interbed (Lite and Grondin, 1988). Anecdotal evidence and well logs indicate that in the OWRD administrative area (fig. 1) groundwater levels are higher in deeper aquifers when they are encountered during drilling. This is evidence of a persistent upward gradient above, and including at least part of, the Frenchman Springs aquifers.

Because deepening of wells is often in response to declining water levels and because most of the deeper wells were installed after groundwater declines began, no reliable estimates of pre-development vertical gradients are available. Moreover, interpretation of the water-level measurements made during drilling is further complicated because water-level data collected during drilling are composite head measurements of units open to the borehole.

Table 1.   Wells where groundwater levels were measured in the Mosier, Oregon, study area, 2005–07.

[**Abbreviations**: USGS, U.S. Geological Survey; OWRD, Oregon Water Resources Department]

Site identification No.	Station name	Measured by	Measurement frequency
453811121212401	02N/12E-19DDD1	USGS	Bimonthly
453838121174801	02N/12E-22ADC1	USGS	Bimonthly
453841121181301	02.00N/12.00E-22BDA01	USGS	Bimonthly
453842121185801	02.00N/12.00E-21ADA01	USGS and OWRD	Bimonthly and quarterly
453845121191401	02.00N/12.00E-21ACA01	USGS	Continuously
453859121223101	02.00N/12.00E-19BBB02	USGS	Bimonthly
453936121210901	02.00N/12.00E-17BCB01	USGS and OWRD	Bimonthly and quarterly
453937121215801	02N/12E-18BDA1	USGS	Continuously
453940121191901	02.00N/12.00E-16ABC01	OWRD	Quarterly
453943121224901	02.00N/11.00E-13AAD01	USGS	Bimonthly
453944121211301	02.00N/12.00E-17BBC01	OWRD	Continuously
453956121205501	02.00N/12.00E-08CDC01	OWRD	Quarterly
454001121244001	02.00N/11.00E-11CDA01	USGS	Bimonthly
454006121214501	02N/12E-07DBD1	USGS	Bimonthly
454010121224001	02.00N/12.00E-07CBC01	USGS	Bimonthly
454011121223901	02.00N/12.00E-07CBB01	USGS	Bimonthly
454013121225901	02.00N/11.00E-12DAB01	USGS and OWRD	Continuously
454013121225902	02.00N/11.00E-12DAB02	USGS	Bimonthly
454015121202701	02.00N/12.00E-08DBA01	USGS	Bimonthly
454020121223401	02N/12E-07BCC1	USGS	Bimonthly
454023121210301	02.00N/12.00E-08BCD01	USGS	Bimonthly
454024121233401	02.00N/11.00E-12BDB01	USGS	Bimonthly
454027121212501	02N/12E-07ADA1	USGS	Continuously
454029121225201	02.00N/11.00E-12ADB01	OWRD	Quarterly
454031121215701	02N/12E-07BDA1	USGS	Bimonthly
454031121224001	02N/11E-12AAD1	USGS	Bimonthly
454032121200001	02.00N/12.00E-09BBC001	USGS	Bimonthly
454032121213101	02N/12E-07AAC2	USGS	Bimonthly
454032121215601	02.00N/12.00E-07BAD01	USGS	Bimonthly
454037121205601	02.00N/12.00E-08BAC01	USGS	Bimonthly
454040121222901	02.00N/12.00E-07BBB01	USGS	Continuously
454043121223801	02.00N/12.00E-07BBB02	USGS	Bimonthly
454046121210501	02N/12E-05CCD1	OWRD	Quarterly
454047121203701	02N/12E-05DCC1	USGS	Continuously
454051121203601	02.00N/12.00E-05DCB01	USGS	Bimonthly
454057121241201	02N/11E-02DDB1	USGS	Bimonthly
454133121204701	02N/12E-05BAA1	USGS	Bimonthly

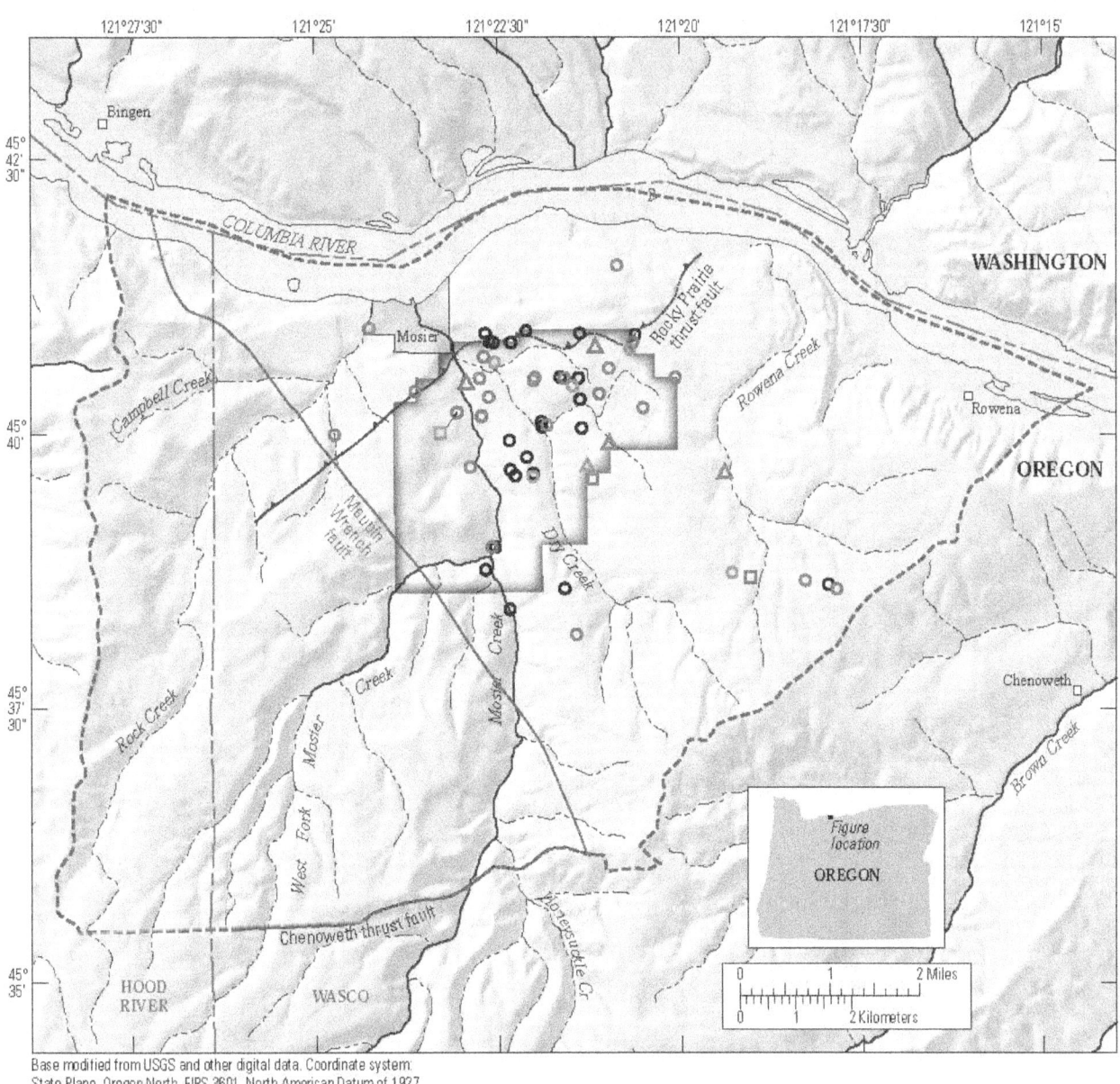

Base modified from USGS and other digital data. Coordinate system:
State Plane, Oregon North, FIPS 3601, North American Datum of 1927.

EXPLANATION

Well measured continuously by

○ USGS

△ OWRD

▢ USGS and OWRD

Well measured periodically by

◉ USGS (bimonthly)

△ OWRD (quarterly)

▣ USGS and OWRD (bimonthly and quarterly)

◉ Well inventoried only (USGS)

▭ OWRD Groundwater Administrative Area

▭ Mosier City limits

- - - - - Geologic model and groundwater-flow model boundary

——— Perennial stream

— · — Ephemeral stream

——— Strike-slip fault
Fault
Strike-slip fault concealed

——▲—— Thrust fault
Fault
Thrust fault concealed

Figure 8. Location of wells where groundwater levels were measured in the Mosier, Oregon, study area, 2005–07.

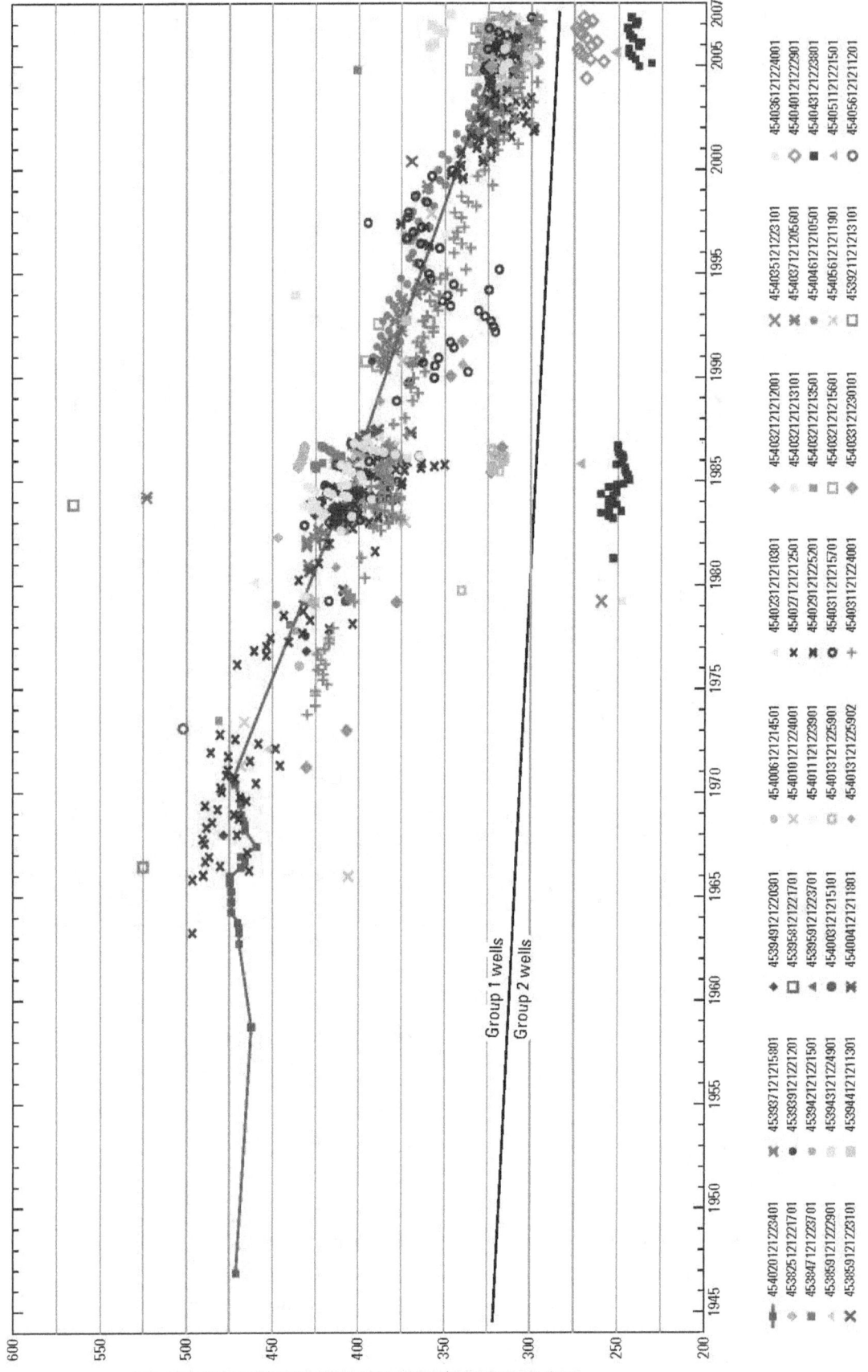

Figure 9. Water levels in selected wells in the Mosier, Oregon, study area, 1944–2008.

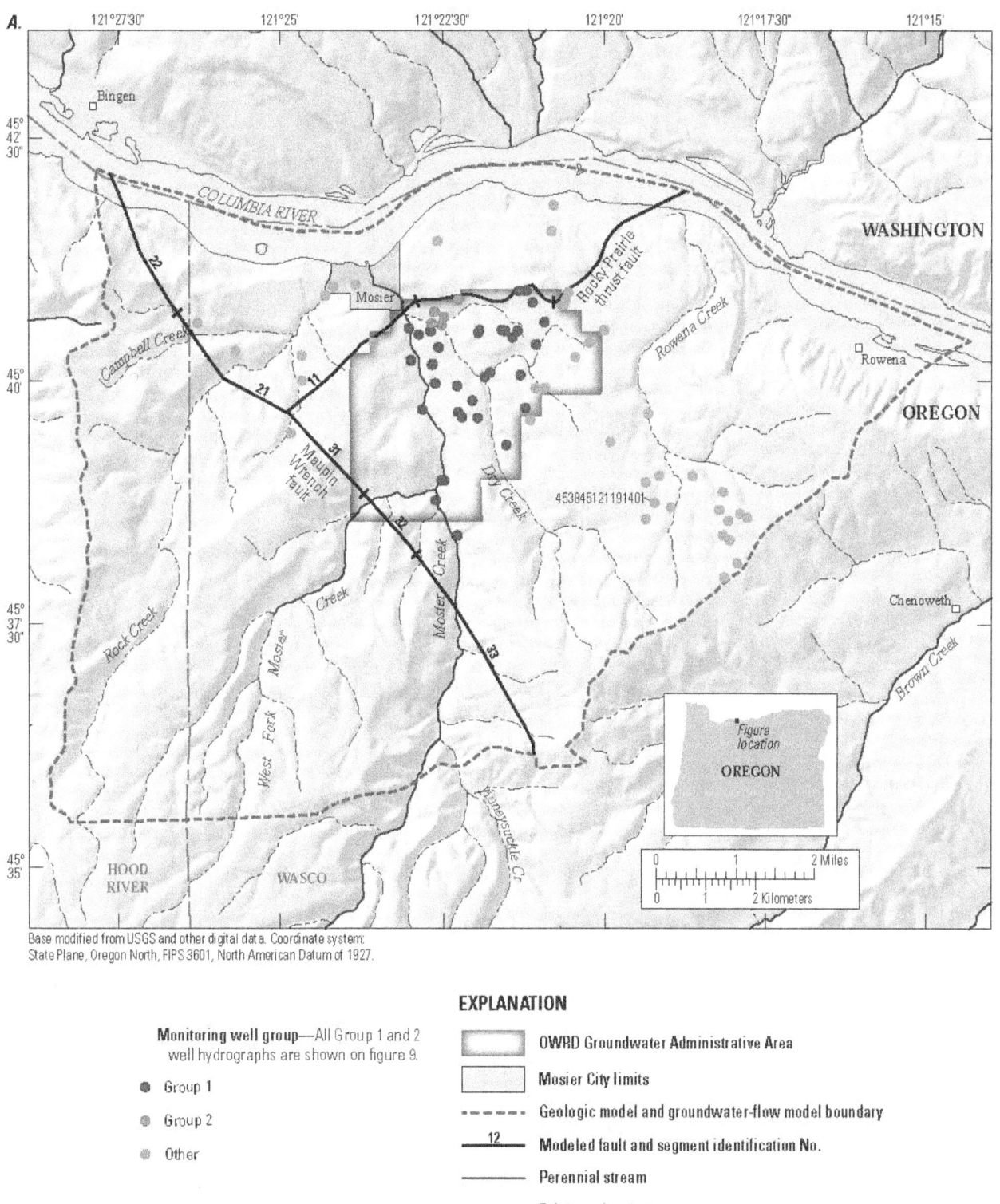

A.

Base modified from USGS and other digital data. Coordinate system:
State Plane, Oregon North, FIPS 3601, North American Datum of 1927.

EXPLANATION

Monitoring well group—All Group 1 and 2
well hydrographs are shown on figure 9.

    ●   Group 1

    ●   Group 2

    ●   Other

                OWRD Groundwater Administrative Area

                Mosier City limits

  – – – –   Geologic model and groundwater-flow model boundary

  <u>12</u>   Modeled fault and segment identification No.

  ————   Perennial stream

  – – – – –   Ephemeral stream

Figure 10. Locations of (*A*) wells exhibiting persistent water-level declines and (*B*) selected monitoring wells in groups 1 and 2, Mosier, Oregon, study area. Water levels in wells are shown in figure 9.

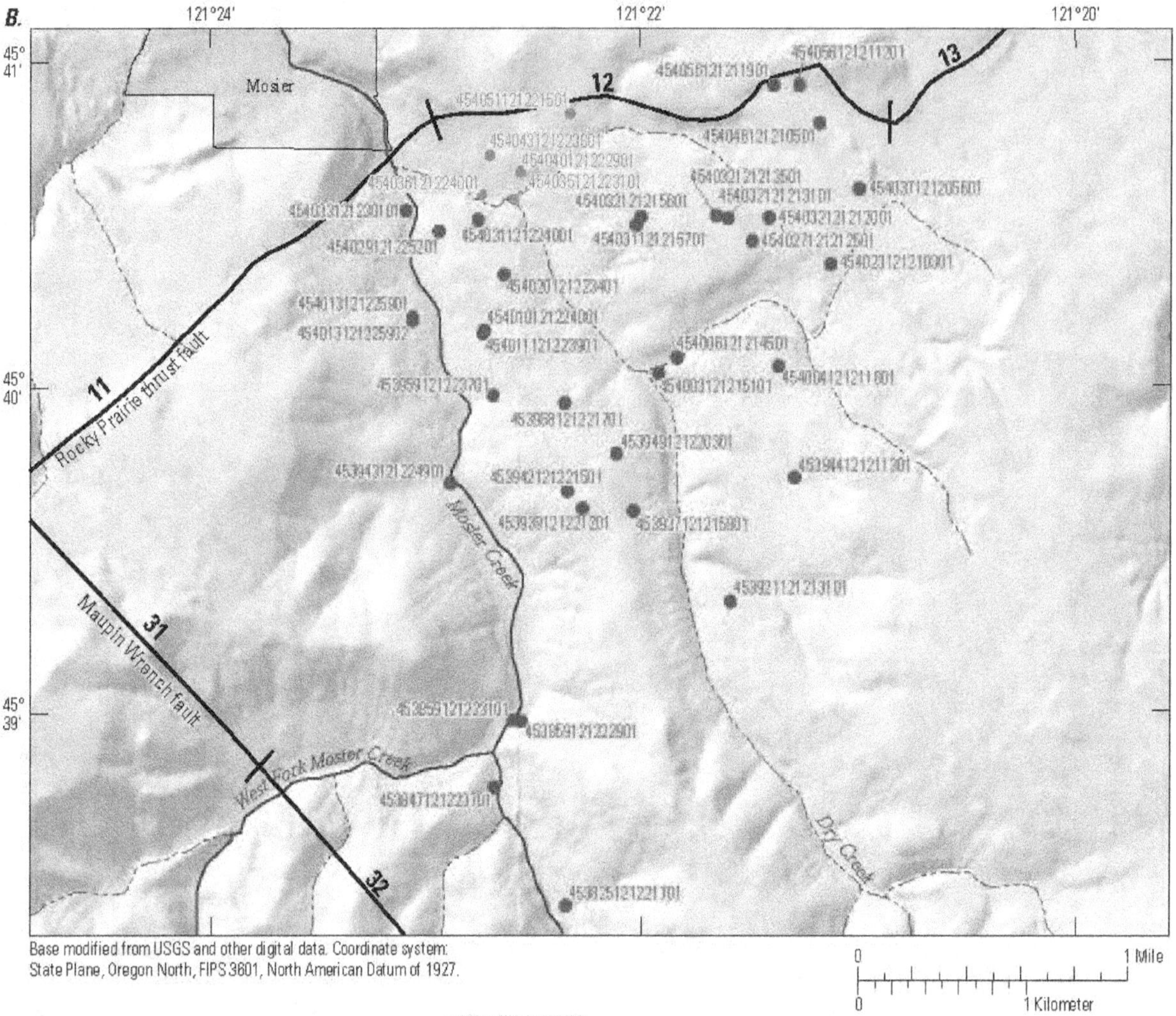

Base modified from USGS and other digital data. Coordinate system:
State Plane, Oregon North, FIPS 3601, North American Datum of 1927.

EXPLANATION

Monitoring well and identification No.—All
Group 1 and 2 well hydrographs are
shown on figure 9.

453847121223701   ● Group 1

454036121224901   ● Group 2

☐ OWRD Groundwater Administrative Area

☐ Mosier City limits

—— Perennial stream

---- Ephemeral stream

<u>12</u> Modeled fault and segment identification No.

Figure 10.—Continued

The vertical hydraulic gradient near the Columbia Hills anticline (fig. 2) is downward. This is the likely source of recharge to the Grande Ronde aquifers. No groundwater-level data are available for the Grande Ronde aquifers in the OWRD groundwater administrative area, and based on geologic modeling results, these aquifers may be connected to the Columbia River to the east, south of the Rocky Prairie thrust fault. The absence of this flow barrier could result in lower groundwater levels in the Grande Ronde aquifers than in the upper CRBG aquifers of the OWRD administrative area, and a resulting downward gradient starting within or below the deeper Frenchman Springs aquifers.

Discharge

Discharge includes all pathways through which water leaves the groundwater system. Groundwater is discharged to surface water features (streams, rivers, springs, and wetlands) as it leaks out of the system or is discharged by pumping from wells. Leaky wells that allow water to flow from one aquifer to another (commingling wells) are internal flow paths, and can affect the rate of discharge into surface water features, but are not considered discharge points from the aquifer system (see section Commingling Wells for a description).

Discharge to Surface Water

Most groundwater in the study area discharges to streams and the Columbia River with the pattern of gaining and losing stream reaches generally controlled by hydraulic compartmentalization of aquifers by geologic faults. When an aquifer is intersected by a stream, groundwater flows into the stream when the hydraulic head in the aquifer is higher than the water level in the stream or river. Groundwater can also flow into streams from springs and wetlands where water is seeping out of the ground above the stream. The amount of streamflow contributed by groundwater is referred to as base flow. Hydraulic head in the aquifers varies over time, providing variable amounts of flow to streams, springs, and wetlands. Flow rates can vary following storms, seasonally, or on longer timescales in response to decadal precipitation patterns or long-term aquifer declines. Because there is low precipitation during summer months, streamflow during this period consists almost entirely of base flow. Average annual base flow is estimated to be approximately 70 percent of total streamflow at the Mosier Creek gaging station for water years (October 1 to September 30) 1964–81 and 2006–07 (appendix C.2).

The spatial distribution of groundwater exchange with study area streams was estimated by measuring streamflow at many points along the Mosier Creek (compare figs. 1 and 11).

The amount of seepage to, or from, a stream reach from the aquifer system is calculated as the difference between the upstream and downstream steamflow after accounting for tributary inflows to and diversions from the reach. Seepage studies of Mosier Creek were conducted in 1962 in a regional groundwater study (Newcomb, 1969), in 1986 as part of a water-availability study by the Oregon Water Resources Department (Lite and Grondin, 1988), and for the current study in 2005 and 2006 (appendix C.2). These latter seepage studies were conducted at various times throughout the year to account for seasonality of water exchange. These data were used during calibration of the PRMS hydrologic model (see section Recharge and appendix B).

Summer estimates of seepage (fig. 11) show persistent groundwater discharge patterns. Flow measurement patterns are complex, consistent with the observation that several aquifers are intersected by Mosier Creek upstream and downstream of the stream gage. The Rocky Prairie thrust fault groundwater-flow barrier is evidenced by increasing streamflow and specific conductance associated with the fault (river mile [RM] 0.8) where groundwater is forced to discharge to the stream. Although there is a pronounced reduction in base flow when comparing the September 1962 streamflow measurements to later measurements, precipitation at the proximal Hood River rain gage (fig. 1) was significantly higher during August and September of 1962 than for the periods preceding all other measurements. For this reason, clear linkages between the declining groundwater levels and base flow cannot be made.

Data from the five seepage studies conducted during 2005–06 (fig. 11 and table C2) provide evidence that the Chenoweth thrust fault (RM 7.1) is also a groundwater-flow barrier. The percentage of streamflow measured at the Mosier Creek gaging station (site 4 on fig. 1) that is in the creek immediately south of the Chenoweth thrust fault (site 1 on fig. 1) ranges between 74.3 and 106.2 percent (table C2), with a median value of 80.0 percent, indicating that groundwater may be forced into Mosier Creek above the Chenoweth thrust fault rather than flowing across the fault through the aquifer system. The single measured value greater than 100 percent (August 2006) indicates that water was lost to the aquifer system below the fault and upstream of the gaging station. The PRMS estimate of groundwater recharge upstream of streamflow measurement site 1 (near the thrust fault, compare figs. 1 and 2), is approximately 16.5 ft^3/s on average for 1955–2007, and the average annual base flow was estimated using the PART hydrograph separation computer program (Rutledge, 1998) to be 20.7 ft^3/s at the Mosier Creek gaging station (streamflow measurement site 4) for 1964–81 (see appendix C.2 for details regarding the use of PART).

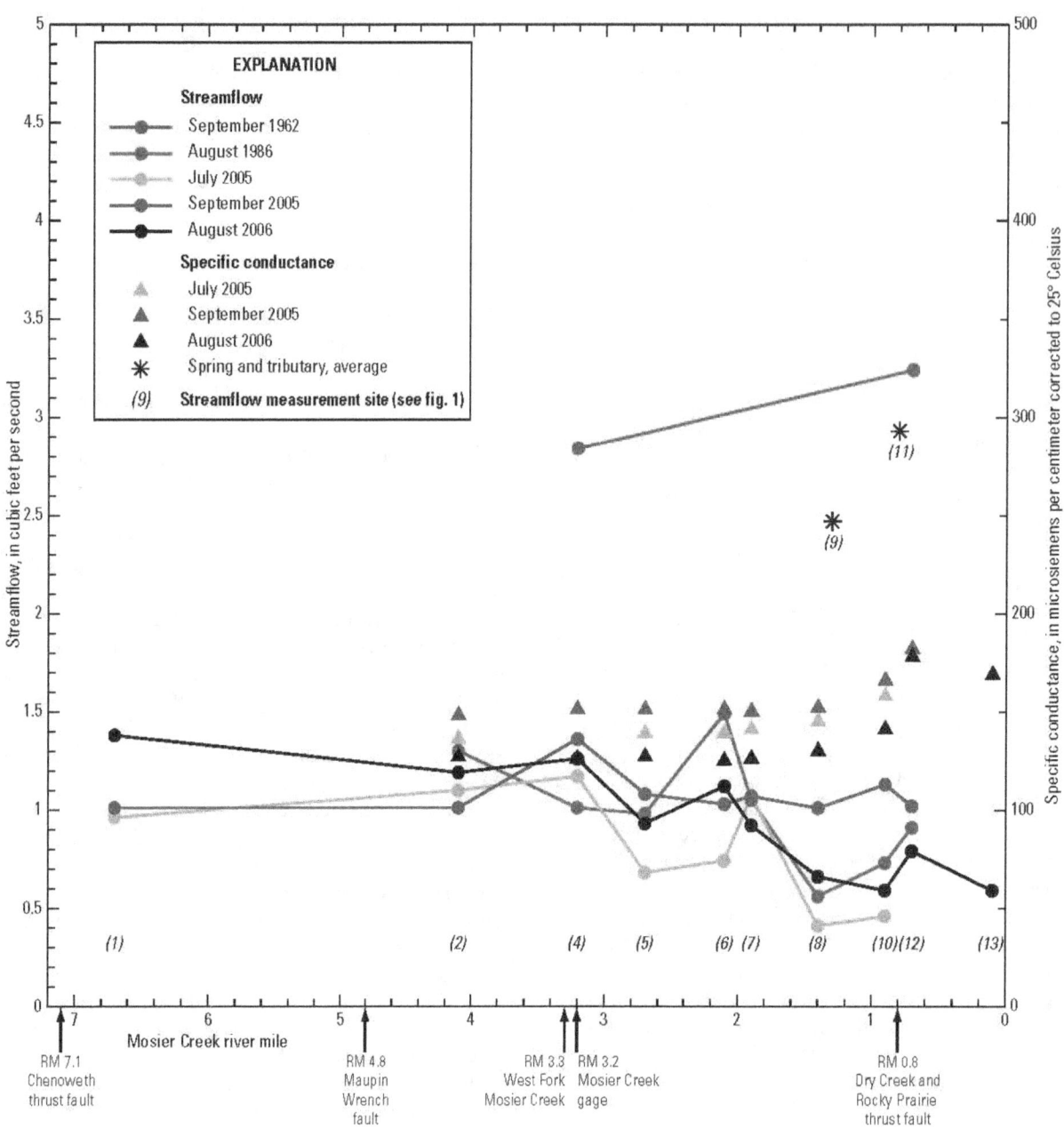

Figure 11.   Streamflow and specific conductance measurements during low flow in 1962, 1986, 2005, and 2006, in the Mosier, Oregon, study area.

If all the PRMS estimated groundwater recharge south of the Chenoweth thrust fault were forced into the stream as base flow above the thrust fault, then it would be 79.8 percent of the estimated annual average base flow at the gaging station, matching the measured streamflow ratios well, providing evidence that the thrust fault may be an effective barrier to groundwater flow.

The control of the thrust faults on Mosier Creek base flow suggests a relation between geologic faults and streamflow that explains the observed flow patterns of Rock, Rowena, and West Fork Mosier Creeks (fig. 2). West Fork Mosier Creek also is a perennial stream with headwaters above the Chenoweth thrust fault and the Maupin wrench fault, indicating that these faults may also promote groundwater discharge to the creek. Even though Rock Creek flows through gravels with no surface expression low in the watershed during the summer, flow was documented above the gravels during all periods of measurement at site 14 (fig. 1). Rock Creek is crossed by several high-angle faults, creating the potential to force groundwater flow into the Creek. If the Chenoweth thrust fault continues to the west beneath the volcanic deposits

(fig. 2), the headwaters of Rock Creek are above this thrust fault, which may result in water from the upper watershed being forced into Rock Creek, similar to Mosier Creek. The ephemeral Rowena Creek is on the eastern side of the study area, receiving less recharge and crossing only one inferred fault, although the creek runs along a mapped fault for some distance (fig. 2). The ephemeral character of Rowena Creek can be explained by the lack of an extensive source area and lack of compartmentalizing faults crossing the creek.

Pumping of Groundwater

In the study area, groundwater is used for irrigation, public supply, and self-supplied domestic uses. Groundwater use began in the first half of the twentieth century, however, most wells were constructed starting in the 1970s (fig. 12). Even though far more self-supplied domestic wells have been drilled, the consumptive use of water in the study area has been primarily irrigation (fig. 13). Estimates of water usage for each category are summarized below and details are discussed in appendix D.

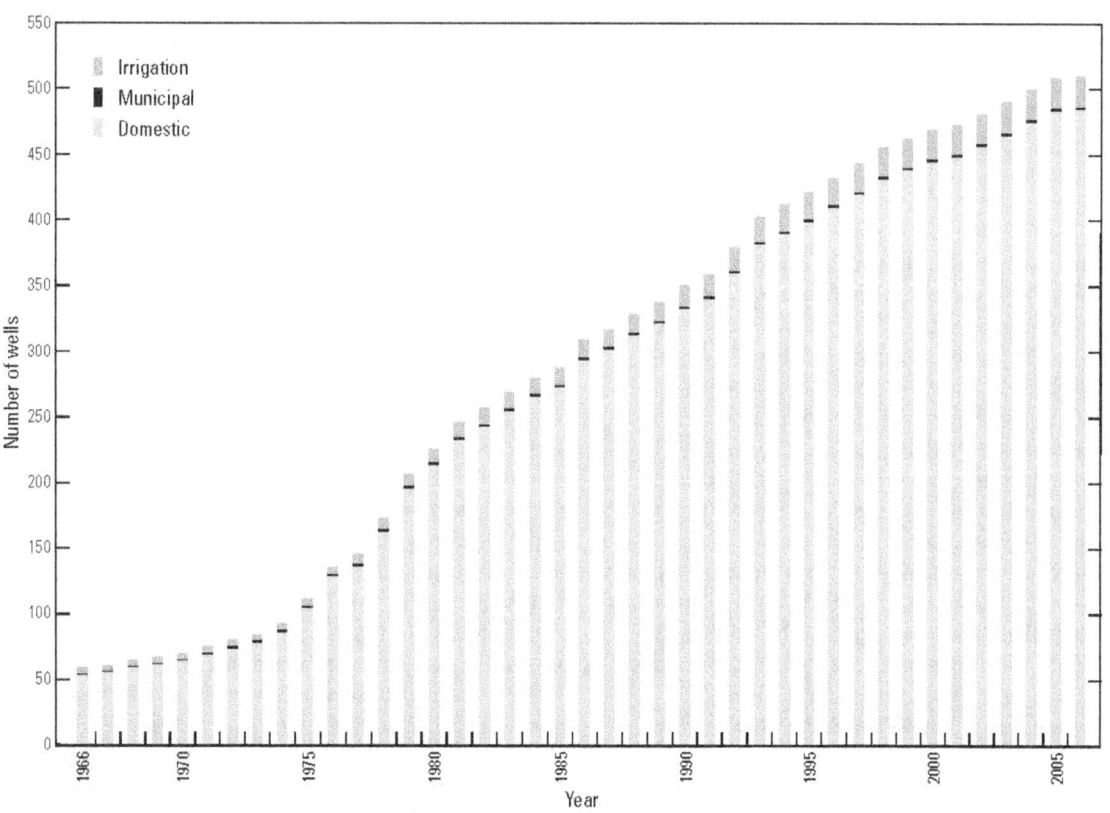

Figure 12. Number of wells and proportion of water-use type in the Mosier, Oregon, study area, 1966–2006.

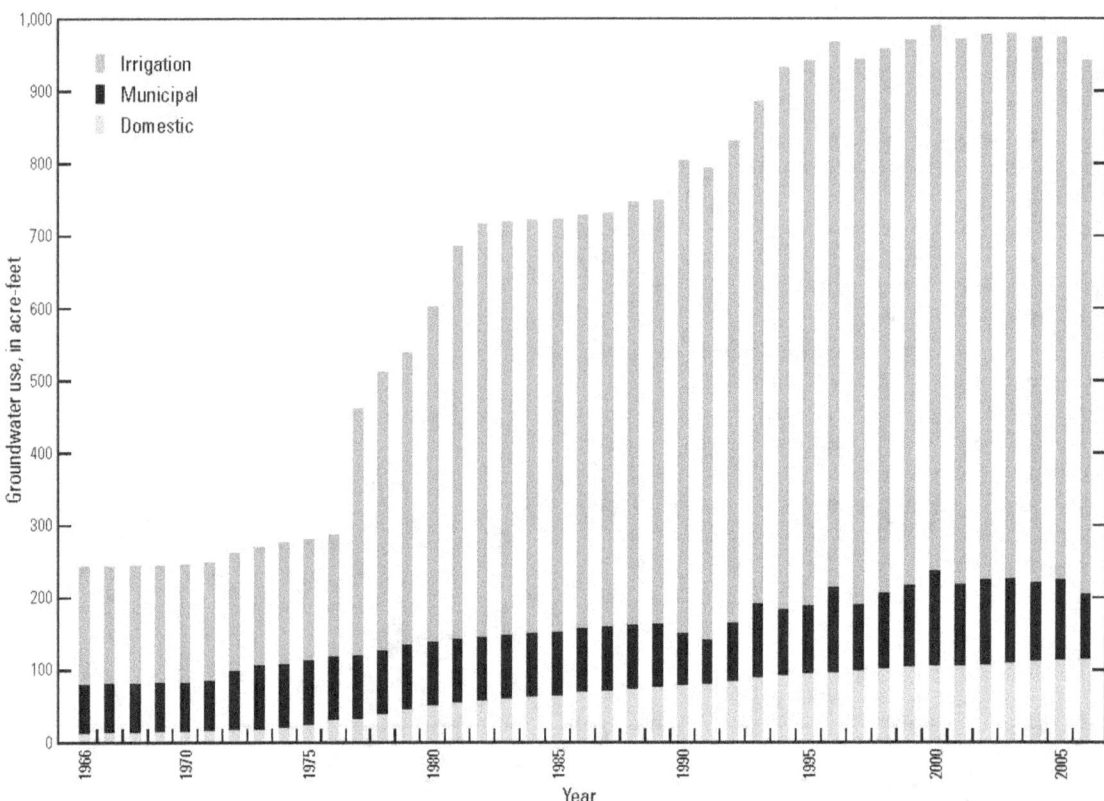

Figure 13. Estimated pumpage and proportion of water pumped for each water-use type in the Mosier, Oregon, study area, 1966–2006.

During the intensively measured 2006 irrigation season, irrigation was the largest use of groundwater, accounting for about 80 percent of total volume pumped, with public-supply and self-supplied domestic accounting for about 10 percent each. A total volume of about 740 acre-ft of groundwater was applied to almost 860 acres from 19 wells in or near the OWRD administrative area (fig. 14). This water was used in the production of fruit tree crops, including cherry and to a lesser extent pear and apple. Wine grapes also are becoming a significant crop in the study area.

Three basic types of irrigation methods are currently used in the study area. In 2006, an estimated 534 acres (62 percent) were equipped with micro spray irrigation, 169 estimated acres (20 percent) were using low efficiency impact sprinklers, and an estimated 155 acres (18 percent) were using drip irrigation. Low-efficiency impact sprinklers were once the standard means of irrigation, but this method is being replaced systematically with methods that are more efficient. The proportion of land using drip irrigation has recently increased.

Public supply for the city of Mosier is another major use of groundwater. The city relies on one primary well to supply water to approximately 430 residents with another well serving as backup water supply (fig. 14). In 2006, the primary well pumped approximately 87 acre-ft of groundwater, and the backup well pumped nearly 3 acre-ft for a combined pumpage of 90 acre-ft. This is about 10 percent of total pumping in the study area. Public-supply water usage from 1989 to 2006 was reported by the city of Mosier. Pre-1989 public-supply water use was estimated using the average 1989 per capita water use rate and estimates of historical population. Details of the city's pumping estimation process are included in appendix D.

In 2006, about 1,200 rural residents pumped an estimated 490 wells, totaling about 114 acre-ft of consumptively used (water used for lawn irrigation etc.) groundwater (about 10 percent of the total pumping in the study area). Non-consumptively used (water used in households) water was assumed to recharge the uppermost aquifer, which is typically the aquifer being pumped by rural residents. Because this indicates no net change in aquifer storage from non-consumptive pumping and recharge (estimated as 60 percent of total annual rural residential pumping), both the non-consumptive pumping and recharge from rural residential wells are neglected in water budgets and model input. Time-varying pumping for rural residential use was estimated based on assumptions about historical population, typical water use per capita, and the percentage of water typically used consumptively (estimated as 40 percent of total annual pumping) (details provided in appendix D).

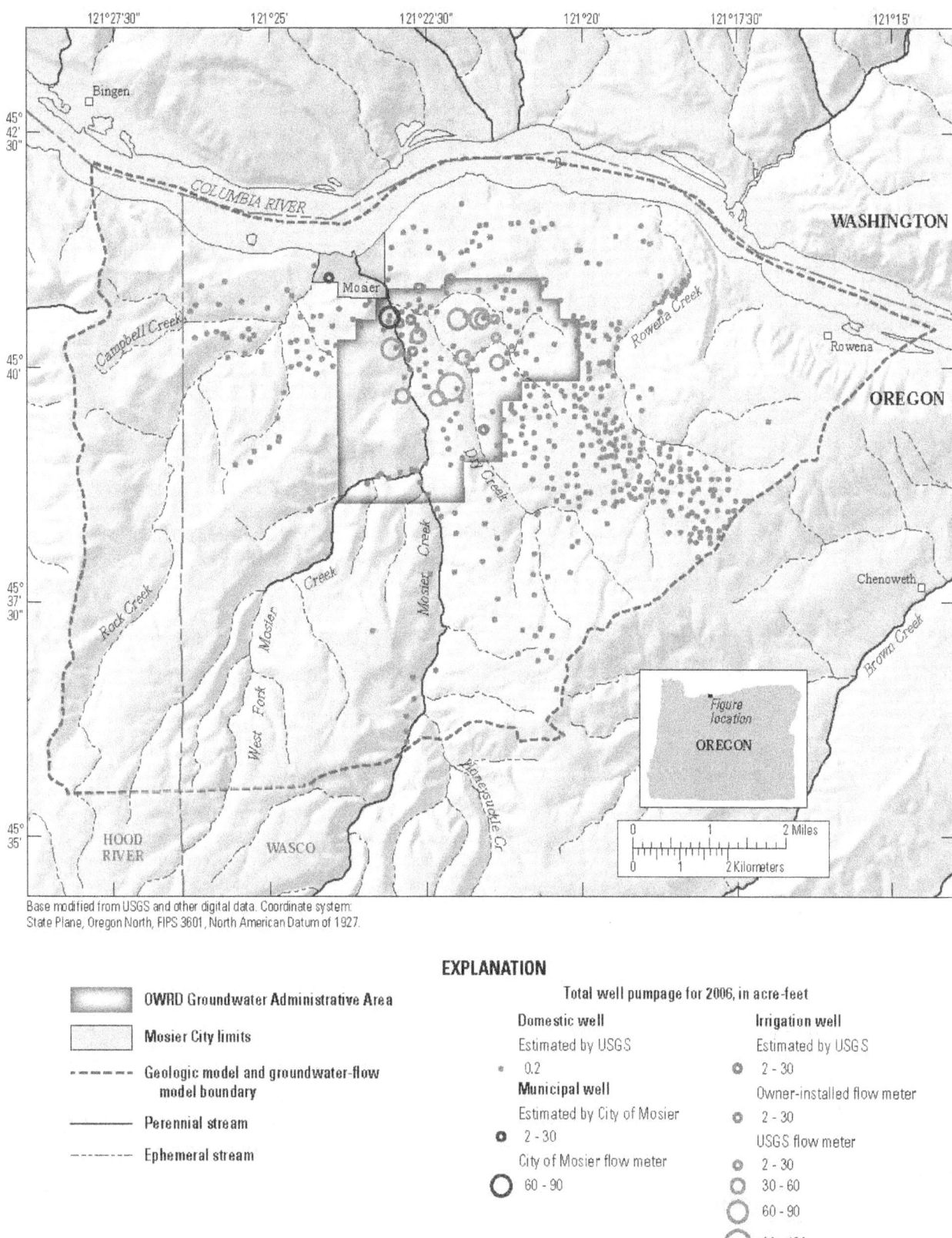

Base modified from USGS and other digital data. Coordinate system:
State Plane, Oregon North, FIPS 3601, North American Datum of 1927.

EXPLANATION

	OWRD Groundwater Administrative Area
	Mosier City limits
– – – – –	Geologic model and groundwater-flow model boundary
————	Perennial stream
– – – – –	Ephemeral stream

Total well pumpage for 2006, in acre-feet

Domestic well
Estimated by USGS
· 0.2

Municipal well
Estimated by City of Mosier
◉ 2 - 30
City of Mosier flow meter
◯ 60 - 90

Irrigation well
Estimated by USGS
◉ 2 - 30
Owner-installed flow meter
◉ 2 - 30
USGS flow meter
◉ 2 - 30
◯ 30 - 60
◯ 60 - 90
◯ 90 - 120

Figure 14. Pumpage estimates by water-use type in the Mosier, Oregon, study area for 2006.

Commingling Wells

Well boreholes drilled through multiple aquifers can allow water to flow between aquifers unless seals are installed to prevent this. Vertical flow through the borehole occurs when there are differences in the hydraulic heads of the aquifers penetrated by the well. Water flows from high hydraulic head to low hydraulic head through the well bore. This mixing (or mingling) of waters from different aquifers provides the name commingling well, which also is sometimes called a cross connecting well.

A number of wells in the study area are documented as being drilled through multiple basalt layers but having a minimum length seal (approximately 20 ft of sanitary seal immediately below land surface) between well casing and the geologic formation. Frequently, wells also only have casing that extends from land surface to the top of the uppermost CRBG unit. Because wells are commonly uncased and open below the top of CRBG, commingling can occur freely between basalt aquifers intersected by the well. Even when casing is installed, if there is no well seal between the casing and formation that prevents flow, commingling occurs in the annular space between the casing and geologic formation (including flow into or out of the overburden aquifer). In the case where aquifers receiving water have a significant ability to retain the water, groundwater levels can increase. However, in the OWRD administrative area, the glaciofluvial overburden and CRBG aquifers have low water storage capacity and are highly transmissive, so most of the water passes through the commingled aquifers into local springs and streams with only a small increase in storage within the receiving aquifers.

Commingling wells allow leakage from the aquifer system that can result in groundwater-level declines. Prior to installation of wells, water levels were higher in the highly permeable CRBG aquifers (fig. 15A) (Lite and Grondin, 1988). Installation of a well with an ineffective seal allows water to flow out of the basalt aquifers into the overburden (fig. 15B). If the overburden is sealed off, then water flows from the deep basalt aquifers into shallow basalt aquifers. In either case, hydraulic heads decline in the deeper aquifers, with the amount of head reduction depending on how easily water flows out.

During geophysical testing of a known commingling well to the south of the Rocky Prairie thrust fault (well 454033121230101, appendix F), the measured upward flow rate through the well ranged from 70 to 135 gal/min (11–22 percent of total annual pumping in the study area). Historically, when aquifer water levels in the deeper basalt aquifers were 150–200 ft higher, and the head contrasts between the deeper and shallower aquifers were higher, this flow rate would have been correspondingly higher.

Possibly commingling wells were identified using a rule-based algorithm for representation in the groundwater-flow simulation model. The probable deepest aquifer was selected by using well depth data and the digital geologic model. Because common practice is to have no casing installed in the length of borehole open to competent basalt, boreholes passing through more than one aquifer were identified as possibly commingling, unless well construction data indicated an effective seal was in place. Rural residential wells with no well depth data were assumed to pump from only the shallowest aquifer and as a result, not commingling any aquifers, but this assumption possibly underestimates the number of commingling wells. To the contrary, the number of commingling wells may have been overestimated because, even though a geologic contact is present in a borehole, productive aquifers do not occur at all locations due to depositional variability of the basalt interflow zones.

Regardless of the possible complicating factors, applying the aforementioned assumptions allowed creation of a reasonable distribution and chronology of well construction (fig. 16) that allows testing of the net effects of commingling wells with a groundwater-flow simulation model. Since approximately 1995, the number of possibly commingling wells has stabilized at about 150. This is presumed to be the result of improved well construction practices in the OWRD administrative area (where most deep irrigation wells exist) and the fact that most new wells are rural residential wells that are typically constructed in shallow aquifers.

Temporal Variation in Groundwater Levels and Changes in Groundwater Storage

Changes in groundwater levels correspond to changes in water storage within the aquifer system. The amount of water in the groundwater system varies in time as a result of hydraulic stresses. Annually, water storage increases during wet winter months as precipitation recharges the system, and decreases during the drier summer months as water continues to discharge from the system into creeks and the Columbia River. Groundwater storage may also vary on longer timescales, such as decadal, resulting from multi-year wet or dry periods. If long-term average groundwater recharge remains the same and no additional water is removed from the aquifer system, groundwater levels will oscillate over time, but the average levels remain constant. This condition is called dynamic equilibrium.

The addition of pumping and commingling wells to the aquifer system has resulted in declines in groundwater storage. These declines are present in many study area wells (fig. 9), and until the aquifer system reaches a new dynamic equilibrium, groundwater levels will continue to decline. The persistent groundwater-level declines are superimposed with seasonal and decadal oscillations, representing the effects of seasonal recharge, pumping, and decadal wet-dry periods.

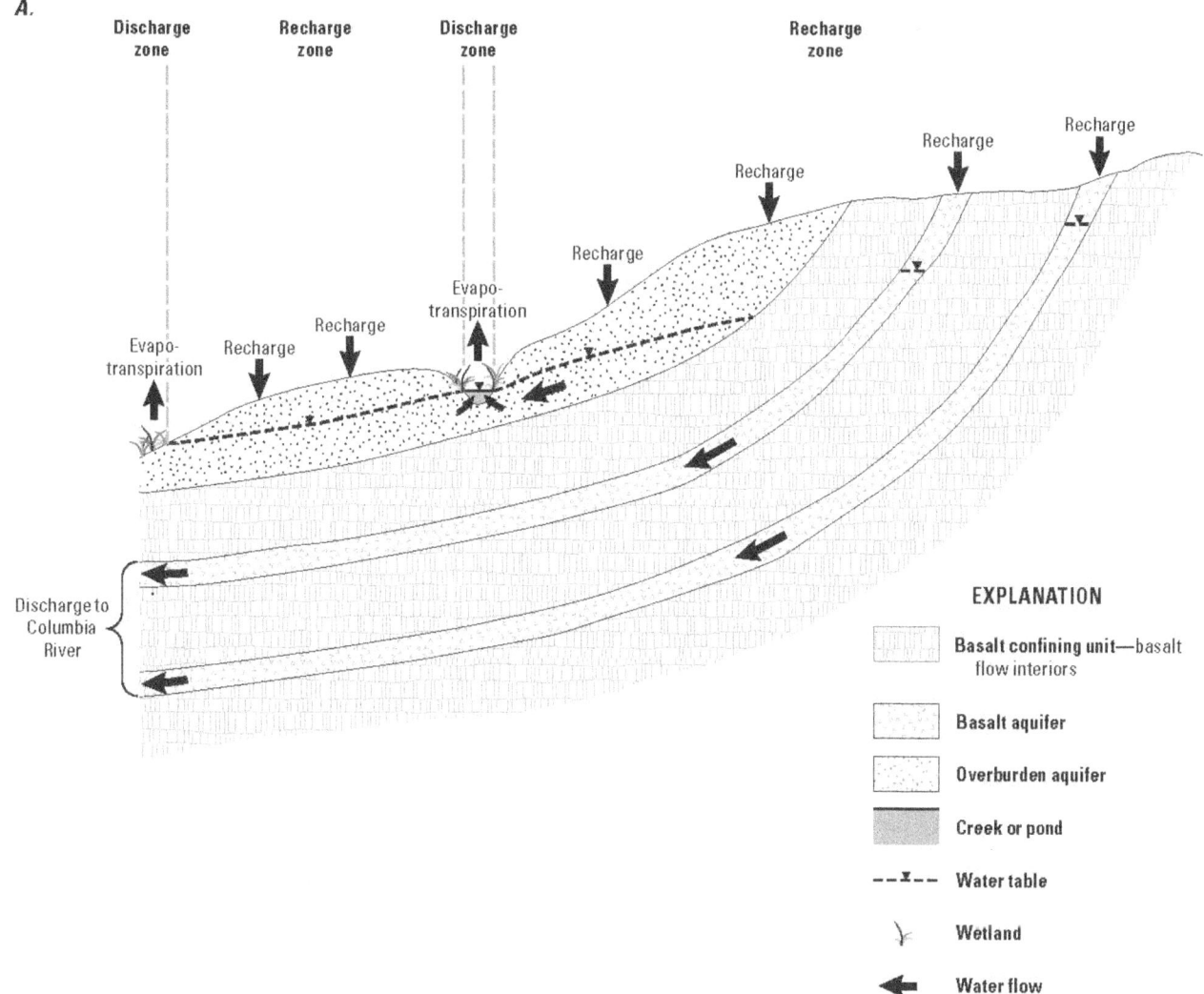

Figure 15. Hypothetical aquifer conditions along a north-south cross-section south of the Rocky Prairie thrust fault in the Mosier, Oregon, study area (A) before development and (B) after installation of a commingling well.

B.

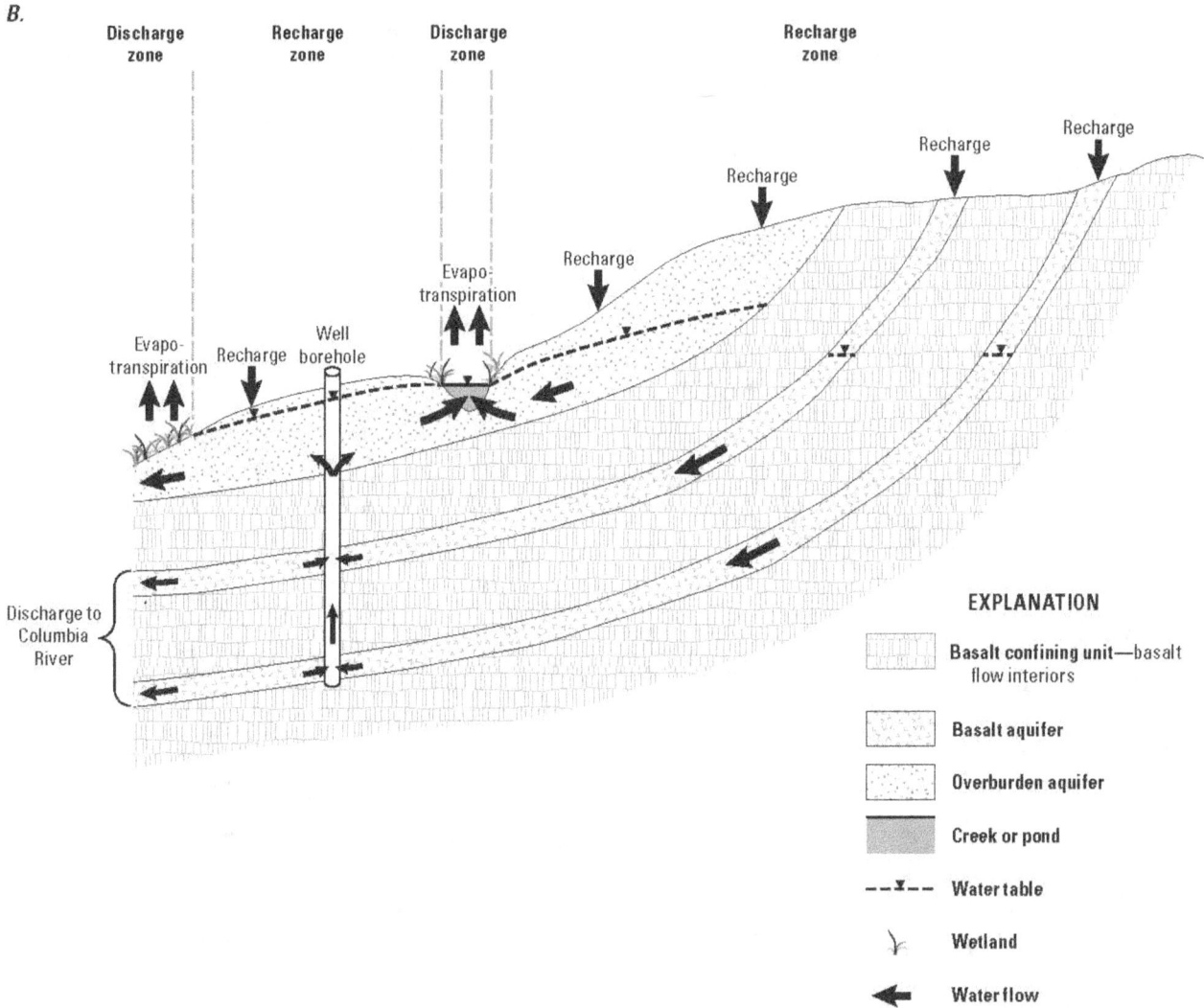

Figure 15.—Continued

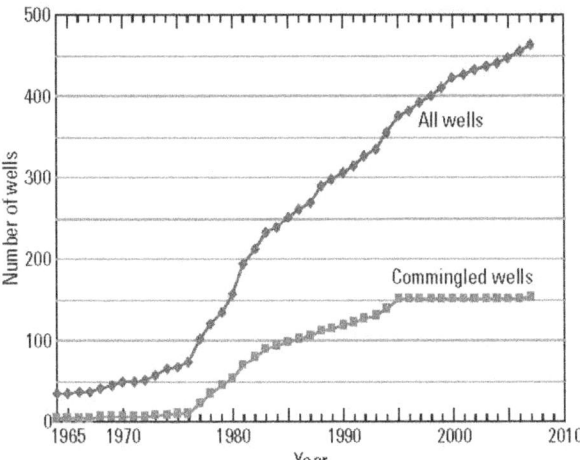

Figure 16. Cumulative number of wells and estimated number of commingling wells drilled in the Mosier, Oregon, study area, 1964–2007. A rule-based algorithm was used to develop a well construction history, including wells that possibly commingle (cross-connect) waters from different aquifers.

Persistent Groundwater Level Declines

Steadily declining water levels in CRBG wells to the south of the Rocky Prairie thrust fult are generally coincident with the OWRD groundwater administrative area and the majority of groundwater pumping in the study area (figs. 9, 10B, and 14). Long-term water level measurements were examined in wells to identify groups of wells with similar hydrologic response (fig. 10A). The largest declines in the study area were measured in Group 1 wells, which have declined at a persistent 4 ft/yr, beginning during the early to mid-1970s. Group 2 wells have a similar response, although the starting water levels are lower, and the rate of decline is smaller (fig. 9).

By 2006, water levels in most Group 1 wells had declined about 150 ft over 35 years with water levels in these wells typically within 25 ft of each other for most of this period. Water levels in several Group 1 wells seemed to be distinctly different when originally drilled (well 454037121205601, for example); however, within 1–2 years, water levels in these wells became similar in magnitude and rate of decline to water levels in other Group 1 wells. Even though the linear decline dominates the pattern of response for Group 1 wells, seasonal and slight interannual trends are apparent, and these variations commonly are reflected in more than one well (fig. 17).

A smaller group of wells with similar behavior (Group 2 in figs. 9 and 10) is clustered immediately to the south of the Rocky Prairie thrust fault. Group 2 wells have lower initial water levels, but these wells are also steadily declining and appear to be trending towards a similar final hydraulic head. Although the thrust fault is mapped to the north of the wells, the part of the aquifer system affected by faulting likely extends towards these wells. Water levels in these wells are interpreted as being driven by the same physical processes as Group 1 wells, but having a reduced response due poor hydraulic connection within the fault-affected zone.

No other well hydrographs have the persistent steep linear declines exhibited by Groups 1 and 2. Water levels in upgradient wells (for example well 453845121191401, fig. 10A) also exhibit declines (fig. 18), although the rate of decline since the mid-1980s is typically smaller than for Group 1 wells. However, comparing the 1978 and 1985 groundwater levels from well 453845121191401 (fig. 18) indicates that water levels dropped about 24.5 feet over 6.5 years, indicating a rate of decline of 4 ft/yr or greater may have occurred during periods since onset of Group 1 declines in the 1970s.

Water-level elevations in all upgradient wells range from a few hundred to more than 1,700 ft higher than Group 1 wells. The data were sufficiently sparse for upgradient wells prior to 1984, that significant groundwater level responses during the 1970s are poorly documented. However, there is a general trend of apparent steeper groundwater level declines in several upgradient wells during the 1970s followed by flattening of the hydrographs in the 1980s. Crude, two-point estimates of average decline during the 1970s are between 4 and 8 ft/yr for upgradient wells.

Seasonal Variation in Groundwater Levels

Water levels in wells fluctuate seasonally in response to changes in recharge, evapotranspiration, groundwater pumping, and streamflow. Beginning in autumn and continuing through mid-spring, water levels rise as recharge from precipitation to the groundwater system exceeds discharge to evapotranspiration, groundwater pumping, and streamflow. From mid-spring until autumn, water levels decline as water drains or is pumped at increased rates from the aquifer system, and during a period when recharge from precipitation is much lower and evapotranspiration is higher.

The seasonal water-level variation ranged from a negligible amount to about 50 ft. Wells with the greatest water-level ranges were in the OWRD administrative area, with smaller seasonal water-level ranges above the administrative area and to the north of the Rocky Prairie thrust fault. Most seasonal water-level changes ranged between 10 and 25 ft in the administrative area (fig. 17).

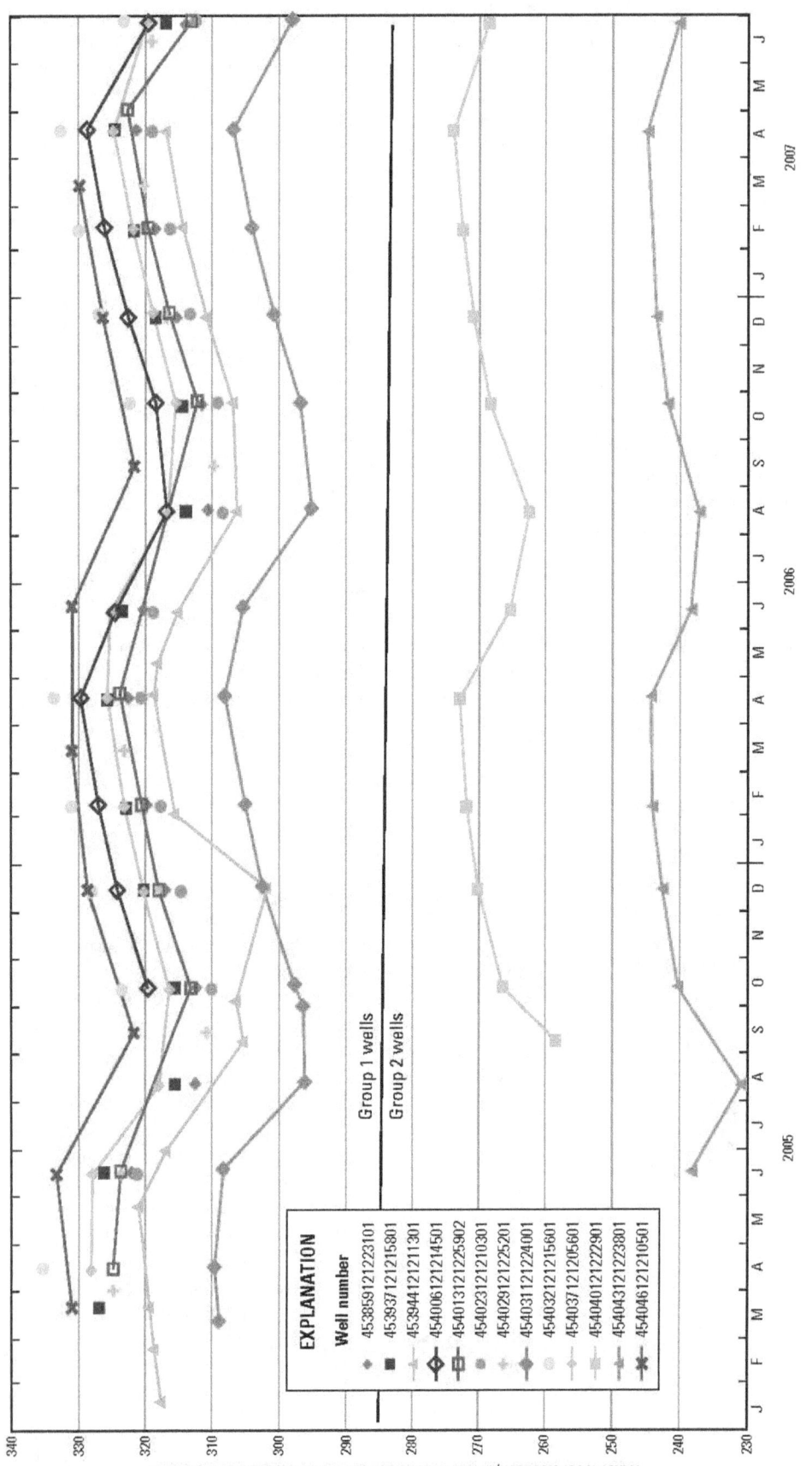

Figure 17.   Selected hydrographs for wells in the Mosier, Oregon, study area.

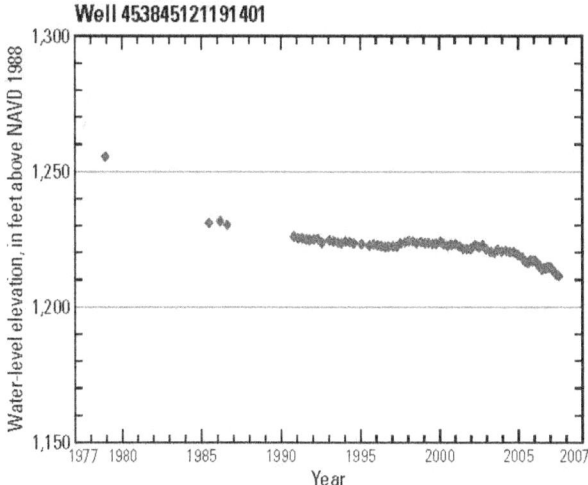

Figure 18. Available data (1977–2007) for well upgradient of the Oregon Water Resources Department groundwater administrative area. Location of well is shown in figure 10A.

Decadal Variations in Groundwater Levels

A comparison between precipitation at Hood River and groundwater levels in the OWRD administrative area wells reveals that a part of the groundwater level changes in the study area is likely related to decadal-scale wet and dry periods (fig. 19). The annual total precipitation by water year is strongly correlated between Hood River to the west and The Dalles to the east, indicating either precipitation record can be used as a surrogate for wet and dry periods in the study area. Average precipitation at Hood River has been approximately 31 inches per water year since 1950, with more persistent decadal-scale wet and dry periods of precipitation after 1970.

To examine the response to decadal variation in precipitation, a shallow and deep pair of Group 1 basalt wells was selected and de-trended. First, the seasonal patterns were removed by selecting late-winter water levels, followed by removal of the linear trend. For both wells, the best-fit slope for the post-1974 data was 3.9 ft/yr of decline. The shallow well (454031121215701) is open to the uppermost basalt aquifer, and water levels follow the wet and dry periods closely (fig. 19). The nearby deeper well (454031121224001) is completed in an aquifer that is several basalt aquifers below the shallow well, and water levels show a decadal scale trend; however, the response is attenuated and lagged by as much

as 10 yrs with respect to the shallow well response. The deep well response is more typical of hydrographs for basalt wells in the OWRD administrative area (fig. 9), indicating the typical climate-driven variation of water levels for wells in the deeper basalt aquifers in this area is approximately 10 ft.

Regularly, the decadal-scale relative high groundwater level of the shallow well corresponds to a decadal-scale relative low of the deep well (fig. 19). When comparing the water-level measurements of these two wells since 1979 (fig. 9), the hydraulic head difference between the well pair has varied between –12 and +67 ft (negative value indicates a downward gradient) with a typical 14 ft upward gradient (computed as the difference between the hydrograph trend lines). The larger variation associated with the shallow well is likely a localized phenomenon, associated with an aquifer of limited areal extent.

Conclusions from Analysis of Groundwater Levels

The following are the primary conclusions from the analysis of groundwater levels:

1. Most basalt wells in the OWRD administrative area have seasonal variations of 10–25 ft, decadal oscillations of approximately 10 ft, and persistent linear declines of about 4 ft/yr. A few wells, likely representing shallow aquifers of limited extent, have larger water level fluctuations, but are still declining at a rate of approximately 4 ft/yr.

2. Groundwater levels outside the administrative area are highly variable, with many exhibiting declines and oscillatory behavior, although documented rates of decline generally are significantly less than 4 ft/yr. Few data are available for most wells upgradient of the OWRD administrative area prior to 1984, although limited data support water-level declines in several wells during the 1970s at rates of 4–8 ft/yr with significantly lower rates of decline after this period.

3. Groundwater levels respond to decadal precipitation patterns. Because the post-1970 period has higher average precipitation than for the 20 years prior, groundwater levels should be rising rather than falling as observed. Therefore, the persistent groundwater declines in the study area cannot be attributed to changes in precipitation.

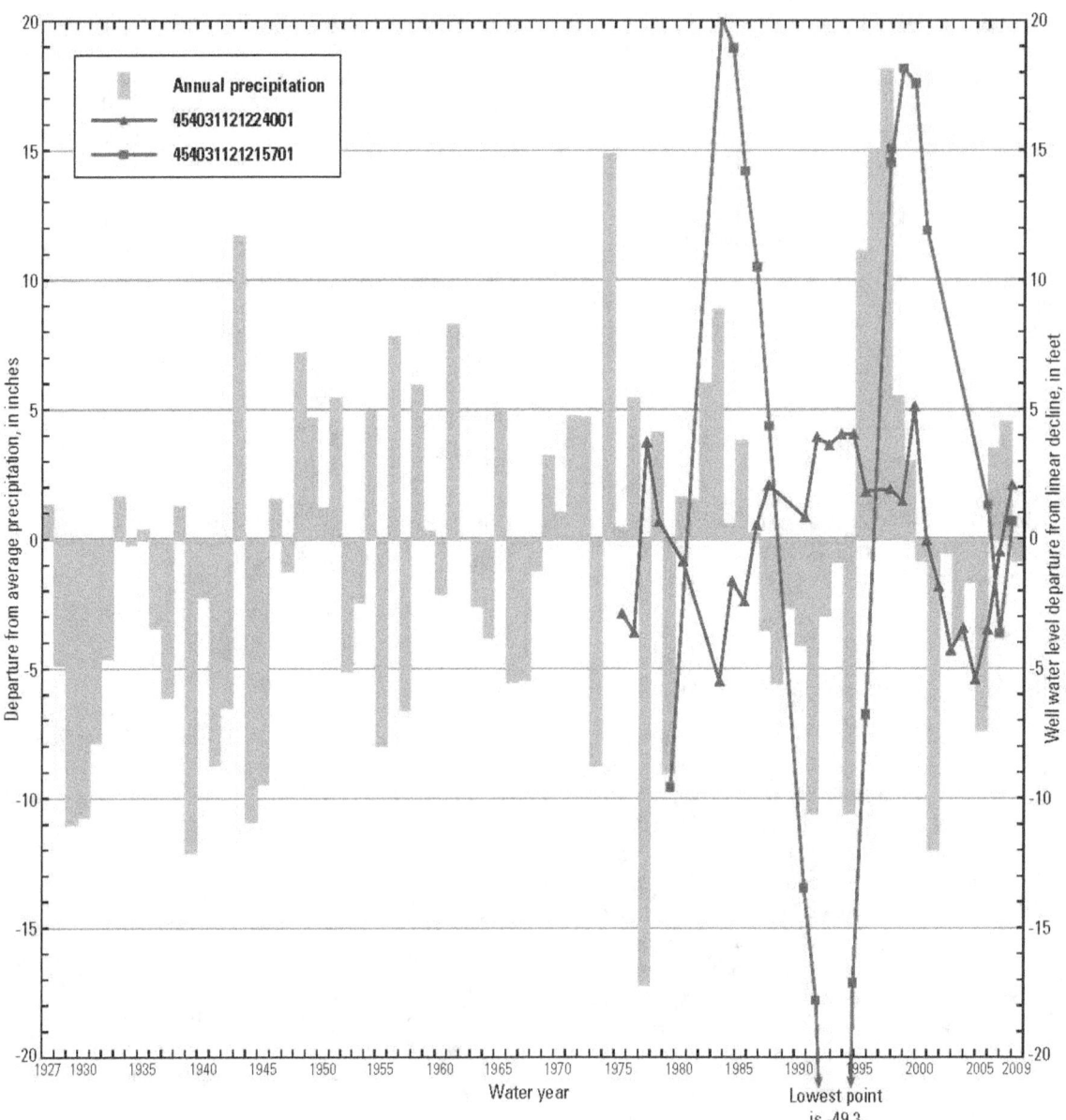

Figure 19.   Decadal response of water levels in Mosier, Oregon, study area wells to precipitation-driven recharge (precipitation measured at nearby Hood River, Oregon). Water levels in a shallow basalt well (454031121215701) show a definite response to wet and dry periods, but most Oregon Water Resources Department administrative area wells (typified by the proximal well 454031121224001) are deeper, showing a lagged, attenuated response. Data for both wells have been de-trended by removing a linear 3.9-foot-per-year decline, and by using water levels representative of late winter to remove seasonal variation.

Conceptual Model of Changes in Groundwater Storage

Changes in groundwater levels are a result of the combination of pumping, commingling, and varying recharge. Reduction in recharge is an unlikely contributor to the persistent groundwater level declines beginning in the 1970s, but the relative contributions of pumping and leakage from the aquifer system due to commingling are more difficult to distinguish. To understand the groundwater conditions and the relative contributions of pumping and commingling to groundwater declines, a groundwater-flow simulation model was developed to incorporate the available data and to represent the complex flow paths within the aquifer system.

The geologic model was used to develop a groundwater-flow simulation model geometry that satisfies the conceptual model of groundwater flow direction (see section "Conceptual Model of the Flow System"). In addition, a conceptual model of storage changes was also developed to aid in the flow simulation model analysis. This conceptual model is illustrated for a single groundwater level in a single basalt aquifer in the OWRD administrative area (fig. 20). Under pre-development conditions, the groundwater system is in dynamic equilibrium (or steady state) with groundwater levels varying seasonally, but exhibiting no long-term trends. This period is represented by the constant groundwater level prior to 1950, after which wells were drilled and pumping begins. Between 1950 and 1972, a few wells were drilled

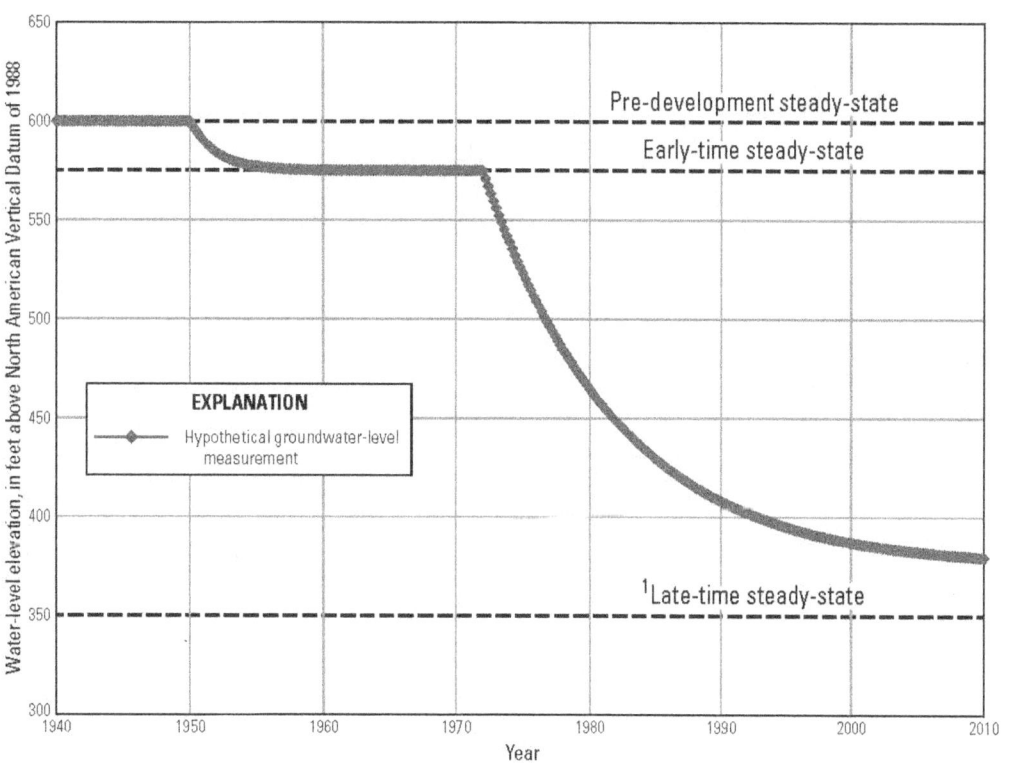

[1]This steady-state value has been selected for illustration only. It does not represent data or modeling results.

Figure 20. The hypothetical water-level response in a single well to groundwater development in the Mosier Creek, Oregon, study area, 1940–2010. After about 1950, a few wells were installed and pumping began, which resulted in a relatively small transient response of system head. The system was apparently in dynamic equilibrium in the early 1970s when installation of additional wells resulted in a larger second decline.

into the upper aquifers in the lower watershed, including the OWRD administrative area. Because only upper aquifers were penetrated, commingling was negligible, and pumping resulted in groundwater declines, with water levels stabilizing at a few tens of feet lower than under pre-development conditions (early-time steady state in figure 20). Starting in the early 1970s, additional wells were installed (fig. 12), increasing the amount of pumping (fig. 13) and the number of aquifers potentially commingled (fig. 16). The resulting groundwater level response to the combined pumping and commingling is much more pronounced, with a significantly lower final water level that has not yet been reached (late-time steady state in figure 20).

Group 1 well water-level data support this conceptual model; although measured groundwater levels are declining at much more linear rate than shown for the conceptual model (compare figs. 9 and 20). Figure 20 emphasizes that if the well configuration, pumping, and recharge remain constant, the system will eventually approach a new equilibrium condition. Under these constant conditions, the rate at which the system approaches the new equilibrium is controlled by the properties governing storage change in the aquifer system, but the final steady-state groundwater levels only depend on the amount of groundwater flowing through the system and the groundwater

flow paths (including pumping and commingling). In other words, the magnitude of the declines provide the most information about the relative effects of pumping and commingling, and the rate of decline provides information about the storage and release of water from the aquifer system. For this reason, the groundwater-flow simulation methods employed to identify the principal causes of the large declines in groundwater levels (see section Separation of Pumping and Commingling Effects) emphasize representation of the magnitude of the declines rather than the rate of the declines.

Practices that would restore groundwater levels are reductions in pumping and repair of commingling wells. If all pumping was ceased in the study area, then water levels will recover (fig. 21). If commingling is negligible, then groundwater levels will recover to pre-development conditions. If commingling is not negligible, then groundwater levels will recover to some lower value. The difference between the recovered value and the pre-development steady-state value can be attributed to the effect of commingling. The difference between the recovered value and the theoretical late-time steady-state value can be attributed to the effect of pumping. This relation formed the foundation of the groundwater-flow simulation model analysis of relative effects of pumping and commingling.

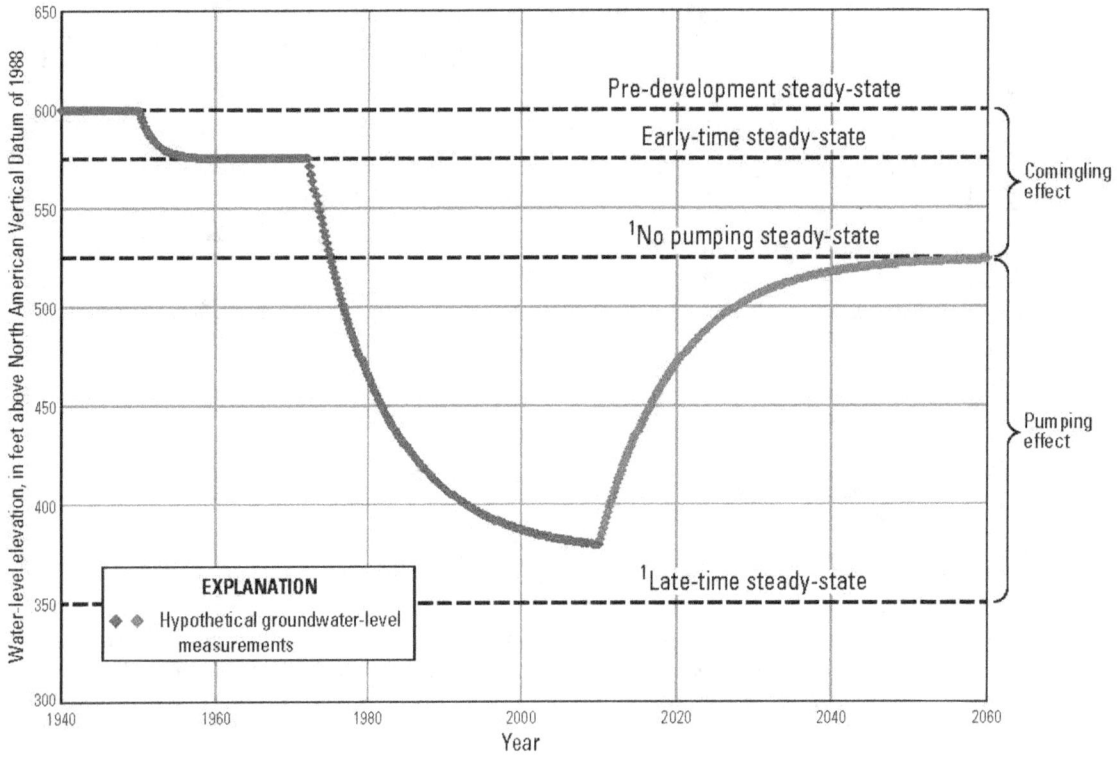

[1]This steady-state value has been selected for illustration only. It does not represent data or modeling results.

Figure 21.   The hypothetical water-level response in a single well to groundwater development in the Mosier, Oregon, study area, followed by the cessation of pumping in 2010.

Groundwater-Flow Simulation

The primary goal of the groundwater-flow simulation analysis was to evaluate the relative contributions of pumping and commingling to the persistent post-1970 declines in groundwater levels in the OWRD administrative area. Following development of a numerical model of groundwater flow using MODFLOW-2000 (Harbaugh and others, 2000) the aquifer system analysis was conducted in three sequential phases using the model-independent parameter estimation software, PEST (Doherty, 2005; Doherty, 2010). At the completion of this analysis, a set of "best estimated" aquifer parameters were selected for presentation in this report, and limitations of the resulting model were identified. The three phases and the principal conclusions of each are:

1. *Rough calibration of a pre-development condition steady-state model* – Initial values of hydraulic parameters were developed from literature values, followed by calibration to the earliest available groundwater level measurements in each aquifer and region of the groundwater model area. Even though there are only two pre-1950 (pre-development) wells with data, it was reasonable to use other early data because documented groundwater declines are relatively small when compared to the range of water-level elevations across the model area (75–1,750 ft). Use of this data allowed a rough calibration of the model, providing reasonable estimates of parameters for use in subsequent analyses. The calibrated model fit the data reasonably well, implying that the conceptual model of flow and the representation of hydrogeology in the groundwater-flow simulation model are reasonable.

2. *Transient modeling of the groundwater-flow system* – A time-varying (transient) version of the MODFLOW model was developed to simulate system hydraulic response during the period of development of water resources. This version of the model performed poorly due to limitations of MODFLOW-2000 when representing the large effect of commingling wells and due to the complex storage characteristics of the aquifer system. To conduct a fully transient analysis, a different groundwater-flow simulation code may need to be used.

3. *Modified transient analysis* – To separate the complex storage problem from the analysis of the relative effects of pumping and commingling, the transient problem was divided into four steady-state simulations to represent the four steady-state conditions shown in figure 2. This formulation of the problem removed the uncertainty associated with estimating the storage terms. Model results indicated that greater than 80 percent of the observed aquifer declines in the OWRD administrative area are attributable to commingling, with the remainder being the result of pumping. A subsequent analysis of water storage mechanisms shows that the long-term linear declines may be the result of draining higher hydraulic head aquifers to supply the confined aquifer system in the administrative area.

The following sections summarize the groundwater-flow simulation model construction (methods used to simulate hydrologic boundaries and assign hydraulic properties to hydrogeologic units), the modified transient analysis and results, and subsequent groundwater-flow simulation analyses designed to aid in selection of management actions intended to restore groundwater levels. Additional details of the simulation analyses are provided in appendix E.

Model Discretization and Boundaries

The groundwater-flow simulation model was created by dividing the model area into 500-ft on a side square model cells of variable thickness and representing springs, streams, rivers, and wells (fig. 22). The thickness of each flow model cell was defined using the hydrogeologic framework, with each layer corresponding to a single hydrogeologic unit in most cases. Model boundary conditions are shown for each layer in figures A9 through A22. The relation between groundwater-flow model layers and the geologic and hydrogeologic layers is shown in figure 3.

Lateral Boundaries

The extent of the model was based on natural hydraulic boundaries. The model area is bounded to the north by the Columbia River, to the south-east and west by ridgelines with adjacent faults, and to the south by a segment of the Chenoweth thrust fault (fig. 2). Water is allowed to flow to the Columbia River, but all other boundaries are considered to be impediments to flow.

Where aquifers are in connection with the Columbia River, a general head boundary allows water to flow to or from the river at a rate proportional to the difference in hydraulic head between the aquifer and the river (fig. 22). Where aquifers are exposed above the river along its shore, a drain boundary allows water to drain out of aquifers with the controlling drainage elevation being the aquifer bottom elevation at the modeled outcrop. This assumption allows simulation of springs along the shore of the river.

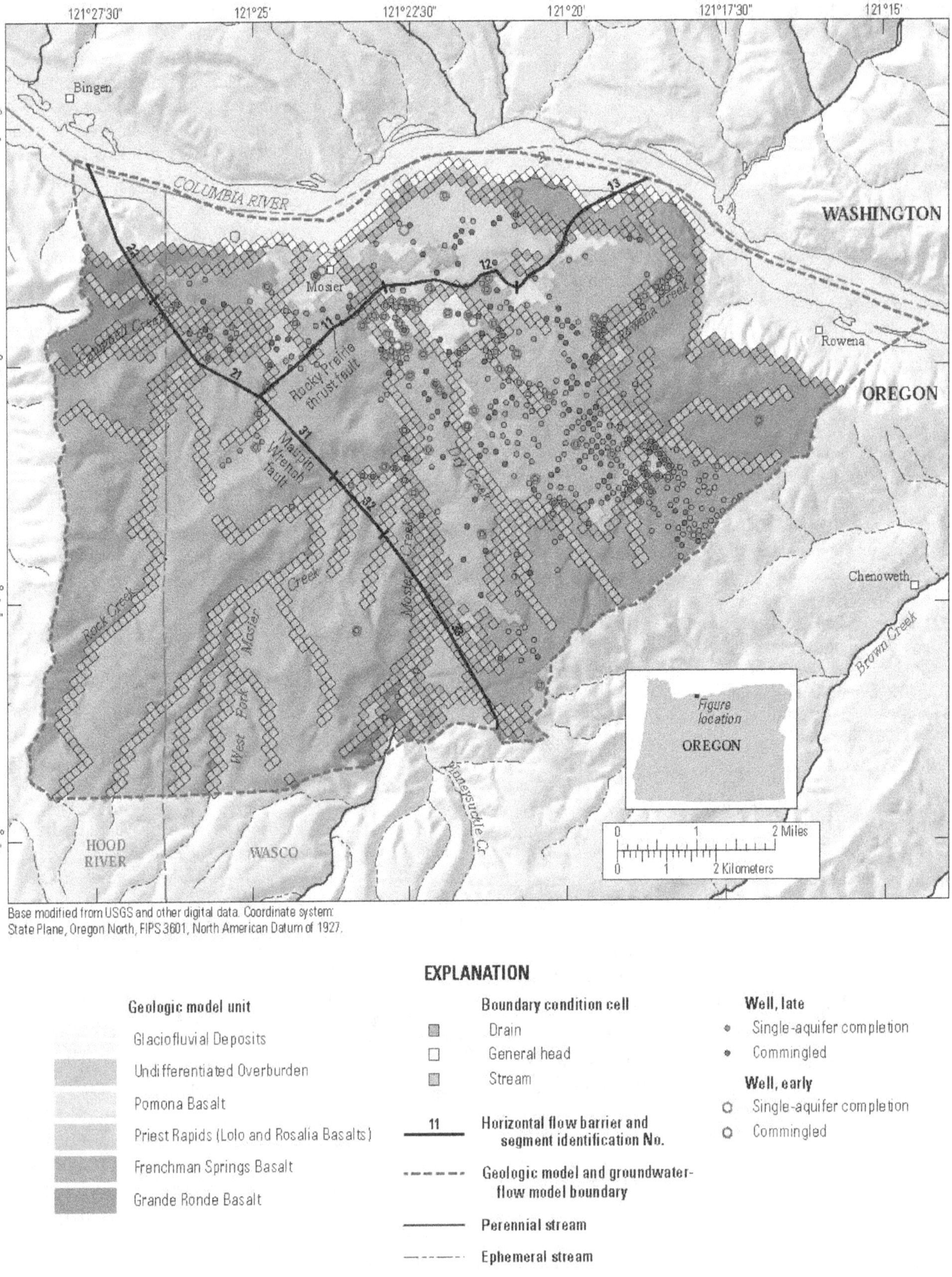
121°27'30"
121°25'
121°22'30"
121°20'
121°17'30"
121°15'
45° 42' 30"
45° 40'
45° 37' 30"
45° 35'
Bingen
COLUMBIA RIVER
WASHINGTON
OREGON
Mosier
Rowena
Chenoweth
Rocky Prairie thrust fault
Mosier Creek
Dry Creek
Rock Creek
West Fork Mosier Creek
Brown Creek
Honeysuckle Cr
HOOD RIVER
WASCO
Figure location
OREGON
0 1 2 Miles
0 1 2 Kilometers
Base modified from USGS and other digital data. Coordinate system: State Plane, Oregon North, FIPS 3601, North American Datum of 1927.
EXPLANATION
Geologic model unit
Glaciofluvial Deposits
Undifferentiated Overburden
Pomona Basalt
Priest Rapids (Lolo and Rosalia Basalts)
Frenchman Springs Basalt
Grande Ronde Basalt
Boundary condition cell
Drain
General head
Stream
11 Horizontal flow barrier and segment identification No.
Geologic model and groundwater-flow model boundary
Perennial stream
Ephemeral stream
Well, late
Single-aquifer completion
Commingled
Well, early
Single-aquifer completion
Commingled

Figure 22. A composite of boundary conditions from all model layers, the model extent, and modeled surficial geology for the groundwater-flow model of the Mosier, Oregon, study area.

The axis of the Columbia Hills anticline forms the ridgetop to the southeast. Immediately to the southeast of this ridgetop is the Chenoweth thrust fault, which is draped down the slope into the next valley. This fold-fault combination is likely to form an effective flow barrier and is simulated using a no-flow boundary condition. Similarly, the western boundary is a ridgetop immediately to the east of a wrench fault of significant offset, also forming a flow barrier and being simulated using a no-flow boundary condition.

The southern boundary corresponds to a mapped extension of the Chenoweth thrust fault system. Two factors support using this feature as a flow model boundary for the groundwater-flow model:

1. The geologic modeling results indicate that many of the upper aquifer units pinch out to the north of the Chenoweth thrust fault (figs. 2, 4, and 5). As a result, most aquifers of interest are not continuous to this point.

2. The Rocky Prairie thrust fault is known to be a major impediment to groundwater flow, so by analogy, the Chenoweth thrust fault is hypothesized to be a flow impediment. Analysis of streamflow data and watershed modeling indicate that most of the groundwater is forced into the stream system above this point, providing evidence that the thrust fault is an effective flow barrier (see Discharge to Surface Water).

Initially, the Chenoweth thrust fault was modeled as a no-flow boundary, with the expectation that later modeling might use a general head boundary to test the effects of flux across this boundary. Results from the modified transient analysis precluded the need for adding additional water at this boundary, because additional water would require that commingling fluxes be even higher. Because commingling was already shown to be the principal cause of declines, the no-flow condition was the conservative case for estimating commingling effects.

Faults

The importance of faults as a control on groundwater flow was evaluated using the groundwater-flow model. In the groundwater-flow model area and below the overburden, the faults with significant offset (potentially juxtaposing aquifers and confining units) which were identified during geologic modeling were simulated as possible impediments to lateral flow through the basalt aquifers (figs. A9 through A22) using the Horizontal Flow Barrier (HFB) package (Hsieh and Freckleton, 1993). Values of conductance of the faults were allowed to vary during calibration and uncertainty analysis. Faults were divided into segments laterally and vertically to test the possibility that water may move more easily through parts of the fault based on fault geometry. For example, the Maupin wrench fault (fig. 2) exhibits high offset

to the southeast, and little or no offset near the Rocky Prairie thrust fault, indicating that the hydraulic conductance will vary along the fault. Similarly, the Rocky Prairie thrust fault can have different properties with depth, especially because the overthrust part of the fault likely only contains Priest Rapids Basalt and younger strata. Fault segment hydraulic conductance values were regularized by indicating that adjacent (vertically or horizontally) segments within the same fault (for example, the Rocky Prairie thrust fault) likely have similar values of hydraulic characteristic (the MODFLOW parameter defining fault conductance).

Streams

In the study area, the aquifers contribute groundwater to streams more efficiently than they gain water from streams. Study area streams lose water to the groundwater system only when the relatively thin aquifers are in direct connection with the stream and stream stage is above the aquifer hydraulic head. If aquifer head is above the stream stage at these locations, then water will flow the other direction, but water may also drain out of the aquifer system into streams through springs and seeps that are above the stream stage. A combination of the MODFLOW drain and stream packages (McDonald and Harbaugh, 1984; Prudic, 1989, respectively) was used to simulate study area creeks. Groundwater-flow simulation model cells containing perennial streams were simulated using both drains and streams, and only drains were used in cells containing ephemeral streams (fig. 22). Details are provided in appendix E.2.

Only two perennial streams originate outside the groundwater model area. These are Rock and Mosier Creeks, originating to the south of the model boundary. Because the reach of Mosier Creek immediately to the north of the Chenoweth thrust fault loses water seasonally, estimated streamflow at the groundwater model boundary was required. This flow was estimated as the total average annual recharge to the drainage area contributing to the stream above the point where it enters the groundwater model area. Inflow at the groundwater-flow simulation model boundary was estimated to be 16.5 ft^3/s for Mosier Creek and 0.911 ft^3/s for Rock Creek. The Mosier Creek estimate is approximately 80 percent of the average annual base flow estimated at the Mosier Creek gaging station (fig. 1), agreeing well with seepage run observations (see section Discharge to Surface Water).

In addition to using drains in model cells intersecting streams, drains were used to represent springs and seeps that would form where aquifers are exposed at land surface if groundwater levels were sufficiently high. Flow from these cells is assumed to flow into streams further downslope. Total groundwater discharge to streams along specified reaches were computed by summing all drain and stream cell fluxes in the contributing drainage area using ZONEBUDGET (Harbaugh, 1990).

Recharge

Recharge estimates were derived from the PRMS watershed modeling results (see Recharge in the Groundwater-Flow System section, appendix B, figure B4). Monthly recharge was estimated for use as input during transient modeling, but the average annual recharge rate was used for the modified transient analysis and all subsequent management scenario simulations. PRMS recharge estimates were made for the entire groundwater model area except for a small relatively low recharge area along the Columbia River (fig. B4). Recharge in this area was estimated as a simple average of recharge for adjacent PRMS model units.

Wells

Several model-area wells are known to commingle the basalt aquifers. Of principal concern to this study is the extent to which vertical intraborehole flow contributes to groundwater-level declines, so representation of the about 500 wells in the model was a critical piece of model formulation. Of these wells, 25 wells account for about 80 percent of the total pumping (see Pumping of Groundwater section). These wells also are typically some of the deepest wells in the area, so accurate representation of the wells is important.

Historical well construction practices in the area have resulted in open boreholes between multiple aquifer zones. Even wells with a casing or liner installed commonly have no functional seal except near the ground surface, allowing hydraulic connections between aquifers through the annular space between the borehole and the casing or liner. The geometry of an important subset of wells was examined in detail, and aquifers pumped and commingled by these wells were assigned to groundwater-flow model units using the geologic model (diagram of this relation shown in figure 3), well logs, and best professional judgment. These wells include all of the high capacity wells and wells installed before and a few years after the declines of the 1970s began. Understanding the geometry of these wells is critical for understanding system response. Groundwater-flow model layers for the remainder of the wells were assigned based on the method described in the Commingling Wells section, providing a reasonable distribution of pumping and commingling for the flow model.

The MODFLOW Multi-Node Well package (Halford and Hanson, 2002) was initially used to simulate commingling wells, but numerical instability of the model resulted. To correct this problem, intraborehole flow between aquifers was simulated by using high vertical conductivity of cells containing commingling wells, and distributing pumping between aquifers based on the transmissivity of the aquifers (see appendix E.2 for details). If the aquifers were not commingled to the extent assumed in the parameterization described previously, then during the calibration process, vertical conductance of well cells was decreased until the net effect of commingling was represented. If commingling was occurring in many wells, the net effect was an increased vertical conductance. In this way, the calibration process revealed the role of commingling wells in controlling study area flow patterns.

Flow and Storage Properties of Hydrogeologic Units

Each hydrogeologic unit was assigned a hydraulic conductivity value, which controls the ease with which water flows through the unit, a specific storage value, which represents how much water is stored under confined-flow conditions due to the compression of water and rock, and a specific yield value, which represents the water that would be released if it were drained under unconfined-flow conditions. These values were assumed to be constant across the study area for each hydrogeologic unit. Additionally, the hydraulic properties are assumed to be similar between units of similar geologic character. For example, CRBG flow interiors are assumed to transmit and store water similarly to each other. Details are provided in appendixes E.3 and E.4.

Groundwater-Flow Model Analyses

The groundwater-flow simulation model was used to evaluate the importance of hydrogeologic controls on groundwater flow and storage. Adjustable parameters were selected for evaluation during the modified transient analysis, and a set of values for these parameters were selected for use in subsequent analyses. The adjustable parameters, selected based on the conceptual model of the system, were horizontal and vertical hydraulic conductivity of all flow model units, hydraulic conductance of the Rocky Prairie thrust fault and the Maupin wrench fault, conductance of streambeds and drains, and precipitation-derived recharge rate.

Parameterization is the process of dividing, grouping, or fixing adjustable parameters based on likely similarities or differences in hydraulic behavior. Tikhonov regularization (Doherty, 2005) was used to represent likely relations between independently adjustable parameters. Parameterization and regularization for major parameter groups include:

1. *Horizontal and vertical hydraulic conductivity of groundwater-flow model units* – Every flow model unit was initially assumed to have isotropic hydraulic conductivity, because CRBG aquifer system horizontal to vertical anisotropy is assumed to result from the contrast between the aquifers and confining units, and these units were separated and represented explicitly in the Mosier area groundwater-flow model. However, the Upper Undifferentiated Overburden confining unit was represented as anisotropic for the final management scenarios analyses to prevent groundwater levels from being simulated above land surface in this unit in the western part of the study area. This was done to prevent bias in estimates of changes in groundwater storage. Conductivity of each model unit was independently adjustable, but regularization was used to create groups with similar values. These groups are the basalt aquifers, basalt confining units, and interbeds. Vertical conductance of commingling well cells was an independent parameter.

2. *Hydraulic conductance of faults* – The Rocky Prairie thrust fault and the Maupin wrench fault were divided into 8 segments in map view and 6 segments vertically (one for each simulated aquifer-confining unit combination, fig. 3), resulting in 48 independently adjustable parameters. Regularization was used to indicate that adjacent fault segments likely have similar properties, while allowing fault conductance to vary with depth or along the fault trace.

3. *Streambed and drain conductance* – Adjustable stream conductance and drain conductance parameters were defined for each groundwater-flow model unit. Regularization was used to create groups with likely similar behavior (for example, drains and streams intersecting basalt aquifers are likely more similar to each other than drains and streams intersecting basalt confining units). During calibration, the conductance of drains located in several low permeability units was tied to the conductance of drains located in the adjacent aquifers, preventing these insensitive parameters from becoming arbitrarily large. This is consistent with the assumption that more water will drain through the aquifers into streams than will drain directly from low-permeability confining units into the stream (though water can drain from confining units into aquifers). Because altering streambed and drain conductance have similar effects on model results, regularization was used to minimize streambed conductance. This regularization condition allows streambed conductance to be sufficiently large to account for stream loss to the aquifer system where data support the need for stream leakage, while allowing drains to account preferentially for most of the base flow contributions.

4. *Precipitation-driven recharge* – Recharge was divided into six hydrogeologic zones (based on surficial geology), each with a parameter to adjust the fraction of PRMS-derived recharge to use as groundwater-flow model input.

Analysis of Persistent Decline of Groundwater Levels

Following poor performance of a fully-transient simulation model, three steady state models were used to simulate the final steady conditions that would result under continued pre-development, early-time (circa 1970), and late-time (circa 2006) stress conditions (fig. 20). This analysis allows examination of the magnitude of declines observed in Group 1 wells, independently of the aquifer-system storage parameters, and is referred to hereafter as the *modified transient analysis*. Model parameters were identical for all three models, with the only difference being configuration of commingling wells and pumping rates (table 2). The three configurations were:

1. Pre-development (prior to 1950)–No wells.

2. Early time (the few years prior to onset of persistent declines)–Pumping rates and commingling well geometries present during 1970.

3. Late time (in the future)–Pumping rates and commingling well geometries during 2006.

Calibration

Calibration is the process of finding reasonable values of model parameters that result in the best model fit to the measurements. Model fit is a measure of how well the simulated values match the measured values (observations). For a complex model with a relatively large amount of data, fit will seldom be perfect, but it is often good enough to allow the model to be useful to gain understanding of system behavior.

Table 2.   Steady state groundwater-flow simulation water budgets for the three configurations used for the modified transient analysis, Mosier, Oregon, study area.

[Recharge from precipitation and total pumping were prescribed model input, and the groundwater exchange with study area streams and the Columbia River were simulated output]

	Pre-development	Early time (circa 1970)	Late time (circa 2006)	Pre-development	Early time (circa 1970)	Late time (circa 2006)	Pre-development	Early time (circa 1970)	Late time (circa 2006)
	(Cubic feet per day)			(Acre-feet per year)			(Percent)		
				In					
Recharge from precipitation[1,2]	3,094,852.0	3,094,852.0	3,094,852.0	25,950	25,950	25,950	98.9	98.9	98.9
Leakage from streams to aquifers	33,515.7	33,004.8	34,096.8	281	277	286	1.1	1.1	1.1
Total	3,128,367.7	3,127,856.8	3,128,948.8	26,231	26,227	26,236	100	100	100
				Out					
Total leakage to streams (drains + streams)	1,329,031.6	1,310,651.3	1,148,866.4	11,144	10,990	9,633	42.5	41.9	36.7
Flow to Columbia River	1,799,365.3	1,788,569.1	1,868,357.3	15,088	14,997	15,666	57.5	57.2	59.7
Total pumping[2]	0.0	28,666.5	111,755.3	0	240	937	0.0	0.9	3.6
Total	3,128,396.9	3,127,886.9	3,128,979.0	26,232	26,227	26,236	100	100	100

[1] Estimated potential recharge from irrigation is significantly less than 1 percent of recharge from precipitation, and it occurs mostly in areas of strong upward hydraulic gradient.

[2] Non-consumptive rural residential pumping is assumed to return to the upper aquifer, the most likely source of this water, through septic systems. As a result, this water is not included in either Total Pumping or Recharge, having a zero net effect on the budget.

The three steady state models were calibrated to groundwater level and streamflow data representative of each period. Because the groundwater system was apparently approaching dynamic steady-state prior to 1970, groundwater-level calibration targets for the pre-development and early-time periods were estimated as the median value of winter groundwater-level measurements during the corresponding period, resulting in 2 pre-development and 12 early-time groundwater-level calibration targets. Winter levels have been collected from many wells historically by the OWRD and are assumed to represent the long-term effects of pumping rather than the seasonal drawdowns associated with summer irrigation. Late-time steady-state groundwater-level data do not exist, although groundwater levels must be at or less than their current levels. Current winter groundwater levels were selected to provide 46 calibration targets, with the understanding that modeled values should be equal to or lower than these estimated values. Calibration target weights were lowered to account for the larger number of late-time targets and the lower confidence that the target values represent true steady-state conditions. Groundwater levels were assumed to represent the hydraulic head at the bottom of each well,

which corresponds to the deepest aquifer intersected. For a large rate of flow between commingled aquifers, effective vertical conductance of well cells is large, and the hydraulic head difference between the commingled aquifers is correspondingly small, which makes the assumption that groundwater levels represent the deepest aquifer a reasonable assumption. The assumption that head differential between aquifers at commingling wells is small was examined at each stage of parameter estimation and prediction, and proved to be reasonable for all sets of parameters that result in a calibrated groundwater-flow model.

Spatially distributed average annual PRMS-derived recharge (see section Recharge for details) was used for all steady-state simulations; 100 percent of PRMS-derived recharge was initially used, resulting in adequate model performance, so these parameters were not adjusted for most of the analyses. Following determination that commingling was the dominant cause of declines, recharge was reduced to 90 percent to evaluate the uncertainty associated with possible errors in estimation of recharge. The commingling effect was smaller but still dominant, indicating study results are valid unless PRMS derived recharge estimates are much too high.

The average annual PART-derived base flow estimate upstream of the Mosier Creek gaging station for the period 1964–81 (see section Discharge to Surface Water for additional detail) was used as a streamflow target for all three steady-state simulations. Because groundwater levels declined during this period, the simulated base flow will be higher than the PART-derived estimate for the predevelopment time period. Similarly, the late-time period simulated base flow will be lower than the PART-derived estimate. PART-derived base flow values were used as calibration targets instead of seepage data because seepage data are available only during the dry season and do not represent the average annual values being simulated (fig. B2). However, simulation results were consistent with spatial patterns of measured seepage.

The modified transient model was calibrated (fig. 23) using PEST (Doherty, 2005), providing reasonable best estimates for model parameters ("Best Estimates for Modified Transient Analysis" in fig. 24). Observations were divided into four groups for use with PEST: pre-development groundwater levels, early-time groundwater levels, late-time groundwater levels, and all stream base flows. Initially, observation weights for groundwater levels for each simulation period were assigned inversely proportional to the number of observations in the group so that periods (groups) with fewer observations had higher weights per observation, providing similar importance to calibration data from each period. This strategy ensured that each groundwater-level observation group (each representing a time-period) provided a non-negligible contribution to the calibration objective function. The stream base-flow group weights were reduced relative to groundwater-level observations to account for measurement unit differences, while ensuring the contribution of this group to the calibration objective function was also non-negligible. This strategy ensured that each observation group provided constraints on model calibration and predictive uncertainty analysis. Pre-development, early-time, and late-time base flow were placed in a single observation group because the PRMS estimate of average annual base flow (1955–2007) was used for all three periods with the expectation that the pre-development and early-time groundwater flow simulation model estimates will be larger than the PRMS value and the late-time estimate will be smaller than the PRMS value.

Adjustment of weights assigned to individual observations within groups was accomplished on a case-by-case basis, with lower weights assigned to observations that were not representative of the processes being examined. Of the 63 measurements and estimates used for calibration, model fit for 2 early-time and about 12 late-time groundwater levels were persistently too low, otherwise fit was good (fig. 23).

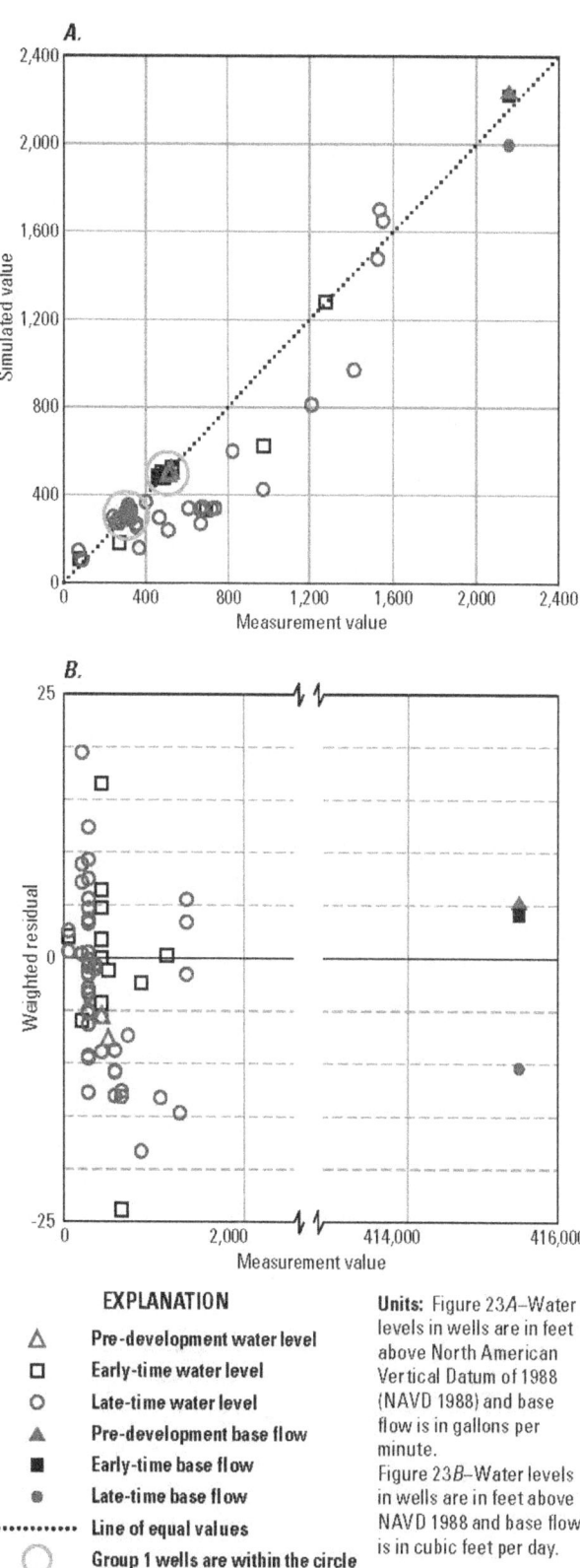

EXPLANATION

△ Pre-development water level
□ Early-time water level
○ Late-time water level
▲ Pre-development base flow
■ Early-time base flow
● Late-time base flow
·········· Line of equal values
◯ Group 1 wells are within the circle

Units: Figure 23A–Water levels in wells are in feet above North American Vertical Datum of 1988 (NAVD 1988) and base flow is in gallons per minute.
Figure 23B–Water levels in wells are in feet above NAVD 1988 and base flow is in cubic feet per day.

Figure 23. Calibration results for the modified transient analysis groundwater-simulation model, Mosier, Oregon. (A) Measurements compared with simulated water levels in wells and base flow, and (B) measurements compared with weighted residuals (simulated value minus measured value).

The reasons for poor fit at these points were examined, and the bias was deemed acceptable for the modeling purposes of analyzing Group 1 groundwater-level declines. The poorly fit groundwater levels were de-emphasized in the automated calibration method by using low observation weights (fig. 23B). Three of the poorly fit points (1 early time and 2 late time) represent water levels in the undifferentiated overburden. Because this unit is thick and highly heterogeneous, the measured groundwater levels are assumed to represent perched groundwater that is recharging the deeper aquifer system. The remaining 11 poorly fit points represent the mid-slope area between the Rocky Prairie thrust fault and the ridgetop (fig. 5A–A'). Calibration with equal observation weights, improved model fit in this area by reducing hydraulic conductivities of the aquifers, but caused excessively large horizontal hydraulic gradients in the OWRD administrative area. This result suggests that there is a conductivity contrast between the upslope and downslope portions of the system. This contrast may be the result of a flow barrier created by a fault or fold, or there may be gradational lateral changes in aquifer hydraulic conductivity created by varying depositional characteristics of the intra-canyon basalt flows. The flow margin of all three upper aquifers is mid-slope (fig. 5A–A').

Because the location and nature of the mid-slope conductivity contrast is not known, representation in the flow model would be uncertain. Rather than developing an uncertain model to represent the conductivity change, the observation weights were reduced for the poorly fit data. De-emphasizing these data is reasonable because (1) the modeling objective is to represent the system behavior in the OWRD administrative area, and (2) the amount of water flowing into the OWRD administrative area is controlled by the prescribed recharge. Streamflow measurements collected for seepage analysis (fig. 11) showed no obvious barriers to flow forcing large amounts of water into Mosier Creek in the area of the likely conductivity contrast (mid-slope), providing evidence that groundwater flow into the OWRD administrative area does not strongly depend on the conductivity contrast.

The resulting calibrated model is appropriate for testing the effects of changes in recharge, pumping, and commingling in the OWRD administrative area. Although the model generally represents aquifer-system behavior, groundwater levels predicted outside of the administrative area are less reliable. The computed groundwater budget is reasonable (table 2) with the model simulating that early time commingling of the upper aquifers resulted in increased streamflow to local streams and reduced flow to the Columbia River. Conversely, for late time, commingling is predicted to increase flow to the Columbia River at the expense of local stream base flow contributions. These results are reasonable given system geometry, but because a predictive uncertainty analysis was not performed, these results are somewhat uncertain, and it is reasonable to assume that combinations of parameters might exist that show reductions in Columbia River base flow for early and late times with corresponding base flow increases to local streams.

Evaluation of Model Parameters

Calibrated parameters match the conceptual model and previously collected data. In addition to the sets of parameters selected as "best" for the previous analysis and for management scenarios (fig. 24), the range of possible parameter combinations was explored during a series of calibrations where starting parameter values and calibration strategy were varied to assess the sensitivity of the estimated parameter values. The analysis was not exhaustive, but did provide confidence that the general values of parameters and relations between parameter values were reasonable. The most sensitive parameters varied between automated calibration runs, with some parameters being highly sensitive at the end of one run but being relatively insensitive at the end of another run.

The CRBG aquifers are the most permeable units, with the younger units generally more permeable than the older units. Lite and Grondin (1988) conducted two one-day pumping tests of the upper aquifers in the OWRD administrative area, estimating transmissivity that is equivalent to hydraulic conductivity values on the order of a few thousand feet per day. These data were used as prior information in the model calibration with low weight, and the calibrated values of hydraulic conductivity for the upper CRBG aquifers were somewhat lower. The lower values are consistent with the observation that upslope aquifer hydraulic conductivity may be lower. Calibrated values of hydraulic conductivity for CRBG flow interiors were 5–6 orders of magnitude less than their associated flow top aquifer values (fig. 24A).

Modeling results indicated that flow barriers associated with faults are important for controlling groundwater flow (fig. 24A). The Rocky Prairie thrust fault consistently had the lowest conductance, with permeability possibly increasing to the east. The Maupin wrench fault was also a barrier with a general trend of higher conductance to the north. Calibrated values of fault conductance indicated that some faults might be less permeable at depth, although trends were not strong.

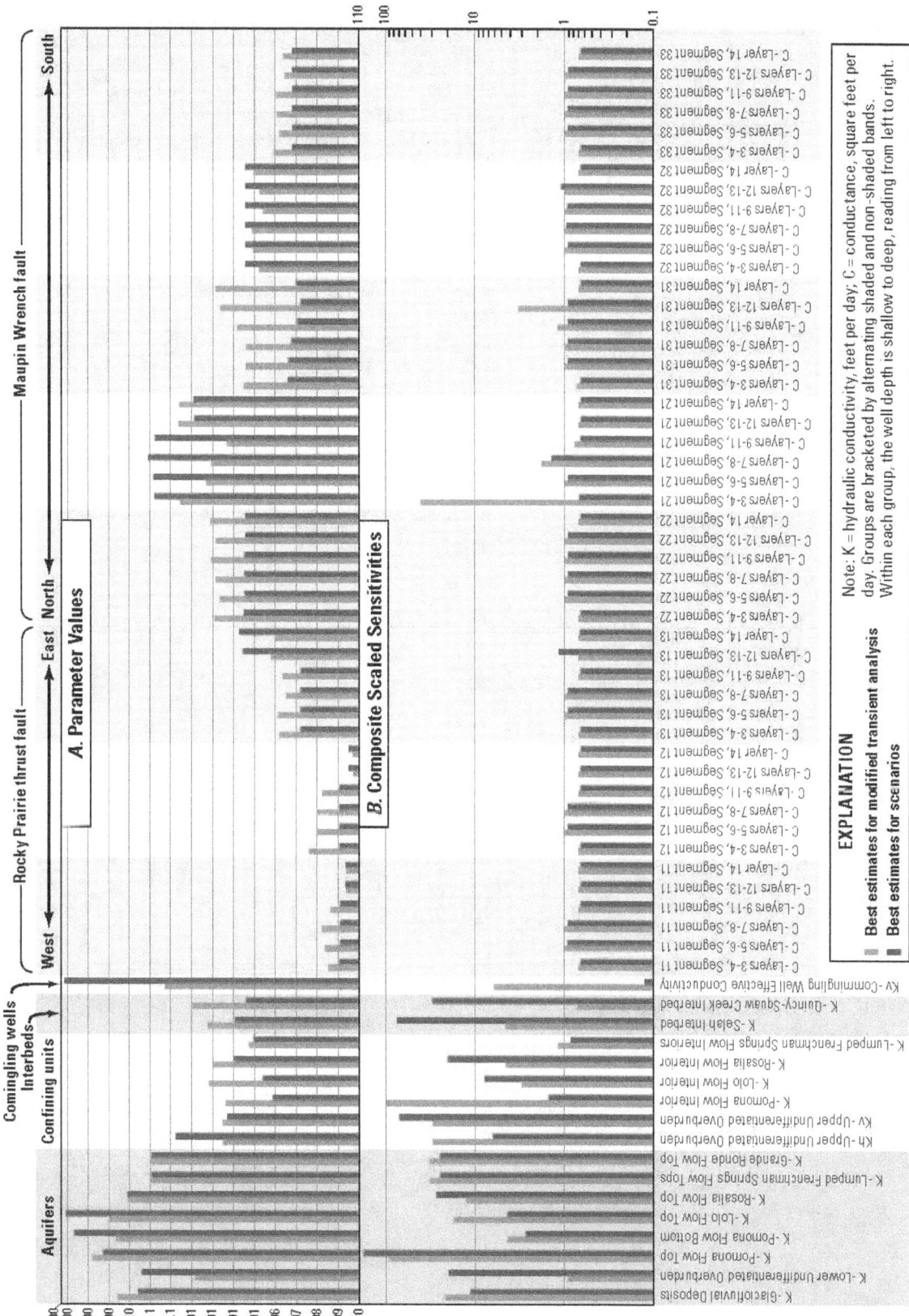

Figure 24. (*A*) Values of adjustable parameters and (*B*) composite sensitivities as computed by PEST (Doherty, 2005). Prior information (with comparatively low weight) used for Tikhonov regularization is included in computation of composite sensitivities.

All of the aquifer and confining unit hydraulic conductivities and fault conductances were sensitive because of the regularization constraints defining the likely relations between units of similar or contrasting properties (fig. 24B). However, the effective conductivity of commingling wells was not always sensitive. When the well conductivity became too large, it became insensitive (compare fig. 24A and B) because, at high values, aquifer hydraulic conductivity limited flow to the commingling wells. The drain and stream conductances for aquifers and overburden had a range of sensitivities (fig. 25), most falling within the range shown in figure 24B. Stream and drain conductances for the low permeability confining units were frequently insensitive, so these were tied to adjacent aquifer stream and drain conductances. Because both stream and drain cells can control the rate at which groundwater leaves the aquifer system, they may act as surrogates for each other during the calibration process. To prevent surrogacy, regularization was used to find the minimum values of stream conductance that result in a calibrated model (appendix E). This allows drain conductance to control the rate of leakage from aquifers to streams, except in reaches where data suggest that aquifers are gaining water from streams.

Separation of Pumping and Commingling Effects

The effects of pumping and commingling were separated by simulating the cessation of all pumping in the model area, but leaving the commingling wells in place. This was accomplished by adding a fourth steady-state simulation to the modified transient analysis model and estimating the groundwater level recovery in all of the Group 1 wells (conceptually shown in fig. 21). Using the calibrated best estimates from the modified transient analysis, the flow model predicted very poor recovery of groundwater levels in the 24 Group 1 observation wells (fig. 26), indicating that commingling may be a large contributor to groundwater-level declines.

PEST was then used in predictive mode (Doherty, 2005) to find the set of hydraulic parameters that still match the data (to a 95-percent confidence, details in appendix E), but that provide the maximum recovery of Group 1 groundwater levels. During this analysis, PEST allows model fit to degrade in order to find the set of parameters that predict the maximum recovery of water levels. The best recovery predicted by the model was typically less than 30 ft for Group 1 wells (fig. 26),

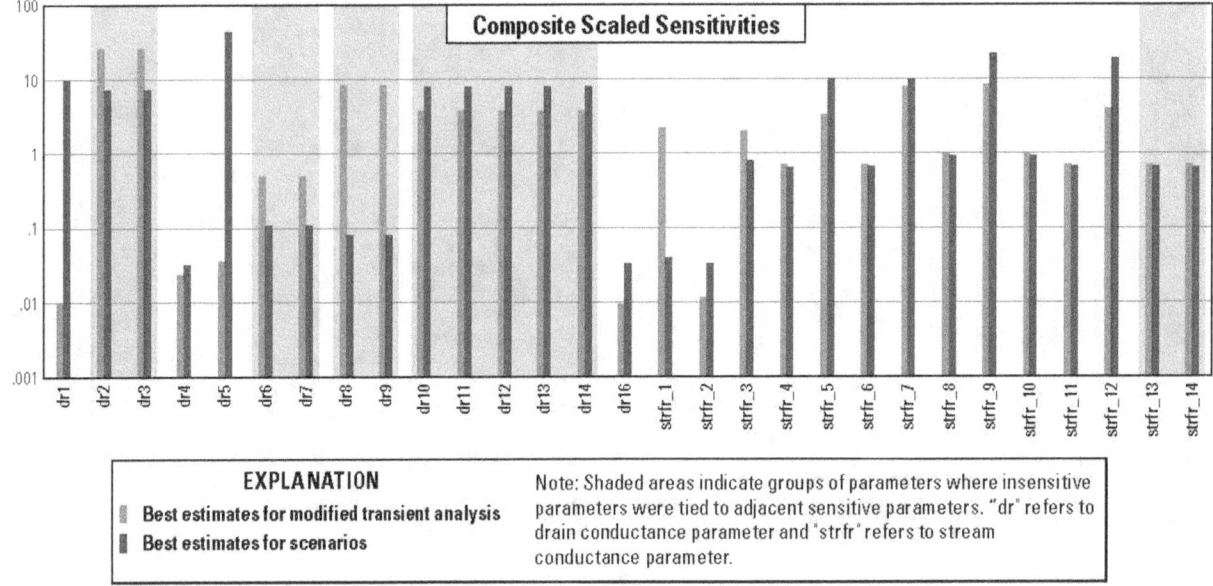

EXPLANATION

Best estimates for modified transient analysis

Best estimates for scenarios

Note: Shaded areas indicate groups of parameters where insensitive parameters were tied to adjacent sensitive parameters. "dr" refers to drain conductance parameter and "strfr" refers to stream conductance parameter.

Figure 25. Composite sensitivities for the stream and drain conductance multipliers as computed by PEST (Doherty, 2005). Prior information (with comparatively low weight) used for Tikhonov regularization is included in computation of composite sensitivities.

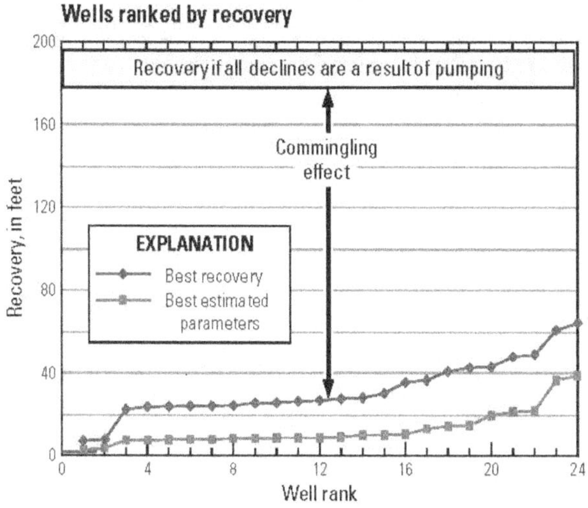

Wells ranked by recovery

Figure 26. Cumulative distribution of simulated recovery that would result from stopping all pumping in the Mosier, Oregon, study area.

with the maximum predicted recovery still accounting for less than one-half the current declines. Some of the parameters attained unlikely values during the analysis, indicating that the recovery predicted by the best-calibrated model ("best estimated parameters" on fig. 26) may be a better estimator. In conclusion, commingling is likely the dominant cause of declines, accounting for greater than 85 percent of the observed declines in Group 1 wells.

The predictive uncertainty analysis above did not consider changes in the zero-flux condition at the Chenoweth thrust fault boundary. However, addition of more water into the aquifer system across this boundary would require that the Maupin wrench fault be less permeable or that the commingling fluxes be even higher to remove additional water. For this reason, the zero-flux condition was a conservative assumption, preventing the need to simulate possible flow at this boundary.

Limitations of the Groundwater-Flow Simulation Model

In addition to the limitations and assumptions described above, the resulting groundwater-flow simulation model and calibration data were examined to identify limitations that may affect future application of the model. This was accomplished by exploring a range of starting parameter values and model calibration strategies. These limitations may be framed in the context of the data required to improve the model for future uses. Six groups of data were identified:

1. There is insufficient data to constrain groundwater-flow conditions to the west of the Maupin wrench fault (fig. 2). In particular, groundwater levels in the thick sedimentary overburden to the southwest, far from the area of current interest, are simulated as being above land surface. This is the result of having only a few groundwater-level measurements for the low-permeability upper undifferentiated confining unit in the eastern part of the study area where data constrain the vertical hydraulic conductivity, which controls the rate at which water is transmitted through the overburden to the aquifers. However, this unit is heterogeneous and stratified, and the effective horizontal hydraulic conductivity controls the rate at which water is transmitted laterally to streams in the western part of the study area (for example, West Fork Mosier Creek (fig. 4). Increasing the horizontal hydraulic conductivity of this unit would allow groundwater levels in this unit to be at or below land surface at all locations.

2. There are insufficient groundwater-level measurements to constrain groundwater-flow conditions through the Rocky Prairie thrust fault and to the Columbia River for the Frenchman Springs and Grande Ronde aquifers. Although the Rocky Prairie thrust fault is a flow barrier to the stratigraphically higher units, the thrust detachment is likely above the Frenchman Springs and Grande Ronde units. However, there may still be deformation of the deep aquifers due to compression that results in restriction of lateral flow. Similarly, there is no data to constrain the connection between these deep units and the Columbia River. As a result, elevated groundwater levels in these deeper aquifers to the southeast of the thrust fault may be explained by poor hydraulic connection across the thrust fault or with the Columbia River.

3. There are insufficient groundwater-level measurements to constrain groundwater-flow conditions in the deep Grande Ronde aquifer system, with only a few groundwater level measurements made in this unit near the crest of the Columbia Hills anticline (fig. 2). Although there is documented and anecdotal evidence that hydraulic-head gradients indicated upward flow in the OWRD administrative area, this evidence is for the upper Frenchman Springs aquifers and above (younger strata), and does not extend to the older, deeper Grande Ronde aquifers. The absence of groundwater-level measurements for the Grande Ronde aquifer in the administrative area, coupled with the absence of Grand Ronde unit exposure in aquifer recharge areas upslope (fig. 2), indicate that not only are groundwater-flow conditions in this unit poorly understood, but that the Grand Ronde may not be a reliable long-term source of groundwater in the study area. Additionally, it is possible that the hydraulic gradients and flow are downward into the Grande Ronde in the OWRD administrative area.

4. There are insufficient groundwater-level measurements to establish the pre-development vertical hydraulic head differences between the aquifers in the OWRD administrative area. As a result, there are nonunique sets of model parameters that can be used to calibrate the groundwater-flow simulation model where the parallel aquifers act as surrogates for each other, transmitting water from the upper to the lower parts of the study area. This generally is accomplished with relatively modest changes in hydraulic conductivity of the confining units. This is because most of the calibration data does not represent the extent to which the aquifers were hydraulically separated before installation of commingling wells. Most groundwater-level measurements represent late-time conditions where the restriction of vertical flow between adjacent aquifers due to the confining units has been greatly diminished by commingling wells.

5. There are insufficient detailed data to determine which wells simulated as commingling are commingling in reality. For this reason, the groundwater-flow simulation model is not an effective tool for evaluating the optimal sequence of well repair.

6. There were too few stream gaging stations with continuous records to allow the use of streamflow data to test hypotheses about changes in spatial distribution of base flow resulting from pumping and commingling. The model predicts a late-time reduction in average annual base flow at the Mosier Creek gaging station (fig. 1) that falls within the noise of the natural variability of the data. Additional streamflow observations at key locations along Mosier Creek likely would have allowed an analysis of changes in base flow upstream and downstream of the current gaging station between the Rocky Prairie and Chenoweth thrust faults. Additionally, continuous streamflow measurements above and below the thrust faults likely would have improved the understanding of the role of these faults as barriers to groundwater flow.

Not all of the previous data limitations are equal for any given future modeling purpose. Of these identified weaknesses, the principal improvement in knowledge necessary to aid in restoration of groundwater levels is the distribution of wells that are commingling in reality.

Evaluation of Potential Management Options

Because commingling likely is the dominant cause of groundwater declines, the groundwater-flow simulation model was used to identify areas most vulnerable to commingling wells and to evaluate combinations of commingling well repair and artificial aquifer recharge that might be used to restore groundwater levels. In all cases, the flow model was used to calculate total change in CRBG aquifer system storage, and the percentage change in CRBG aquifer storage was used as the metric of comparison. Because this performance measure was selected, assumptions were made to address some of the limitations of the groundwater-flow simulation model (see section Limitations of the Groundwater-Flow Simulation Model). These assumptions resulted in two additional constraints and one additional adjustable parameter, and the model was recalibrated to estimate a new set of "best" model parameters for use in evaluating management scenarios. The constraints reflect anecdotal information, best professional judgment, and information gained during the modified transient model analysis. Unlike for the modified transient analysis of commingling and pumping, a predictive uncertainty analysis was not performed. As a result, estimates of change in storage should be used only for comparison between scenarios and not to make absolute estimates of expected groundwater-level recovery at any single point.

The first additional constraint is that pre-development water levels in the overburden must be at or below land surface. This constraint was added because groundwater levels above land surface were simulated near the western boundary of the study area frequently during the calibration process, and these simulated levels are likely erroneous. High simulated groundwater levels in this area probably result from the modeling assumption that the upper undifferentiated overburden is isotropic (despite its high heterogeneity) and from the lack of calibration data to the west to constrain simulated groundwater levels in the overburden to below land surface. The additional constraint was imposed to prevent erroneous groundwater levels in this area from unduly biasing computation of changes in total basalt aquifer storage.

The previous calibrated values of hydraulic conductivity for the undifferentiated overburden unit were controlled by the vertical permeability of this unit near the OWRD administrative area where the unit is less extensive but few calibration data exist. In this area, the primary function of the overburden is to transmit recharge into the deeper aquifer system. To the west, in the trough created by the Mosier syncline (fig. 2), the Mt. Hood volcaniclastic deposits are likely stratified with alternating sequences of more and less permeable materials. In this area, the permeable layers are likely important for controlling lateral flow to several creeks (fig. 2). To allow the groundwater-flow model calibration process to concurrently simulate lower groundwater levels in the undifferentiated overburden to the west and match calibration data near the OWRD management area, the upper undifferentiated overburden-confining unit was simulated as anisotropic, adding one additional adjustable parameter to the calibration process.

The second additional model calibration constraint assumes an almost uniform upward gradient in the basalt aquifers in the OWRD administrative area. Because commingling is significant, the Lite and Grondin (1988) upward hydraulic head gradient of about 70 ft across the Selah Interbed is assumed to be a lower bound for the pre-development hydraulic head differential. Anecdotal evidence suggests a significant upward gradient was encountered each time wells were drilled into deeper aquifers in the OWRD administrative area, so a pre-development head difference across each confining unit was assumed to be 100 ft. This estimate is uncertain because virtually all measurements made below the Pomona basalt aquifers have been affected by commingling of the aquifers. Previous calibrations that disregarded the anecdotal evidence resulted in some calibrated models where hydraulic heads between aquifers were smaller than anecdotal evidence supports. This effect likely is an artifact of the sparse pre-development and early time data, and is not a reflection of the model's ability to replicate this behavior. Inclusion of this second constraint (artificially) rectifies this problem. The uniform upward gradient constraint was enforced by adding a vertical string of artificial observations, one in each basalt aquifer above and including the Frenchman Springs aquifer, immediately to the southeast of the Rocky Prairie thrust fault in the OWRD administrative area. The 100 ft head difference across each confining unit was weighted equally large for the upper three units and 4 orders of magnitude smaller for the confining unit between the Rosalia and Frenchman Springs aquifers. No head difference condition was applied between the Frenchman Springs and Grande Ronde aquifers.

Even though these additional constraints and parameters may improve the groundwater-flow simulation model for many purposes, the previous model (without these additional constraints) is more conservative for evaluating the roles of commingling and pumping. This is true because the extra parameter (a possible new degree of freedom for the predictive analysis) and head observation constraining the upper undifferentiated overburden are largely independent of the declines in the OWRD administrative area (indicating neither the degree of freedom nor the observations constrain the predictive analysis), but the uniform upward gradient requirement (extra observations that constrain the predictive analysis) narrows the range of acceptable models that match the system response (see section Establishing Confidence Intervals for Predictions in appendix E.4). Because the previous model was conservative, the analysis separating commingling from pumping effects need not be repeated for the model with the new constraints.

When the additional constraints and parameter were added to the groundwater-flow simulation model, the model calibrated easily with comparable error to the previous best calibration (compare fig. 27 to fig. 23A). The resulting parameter values (fig. 24) were used for all management scenario analyses.

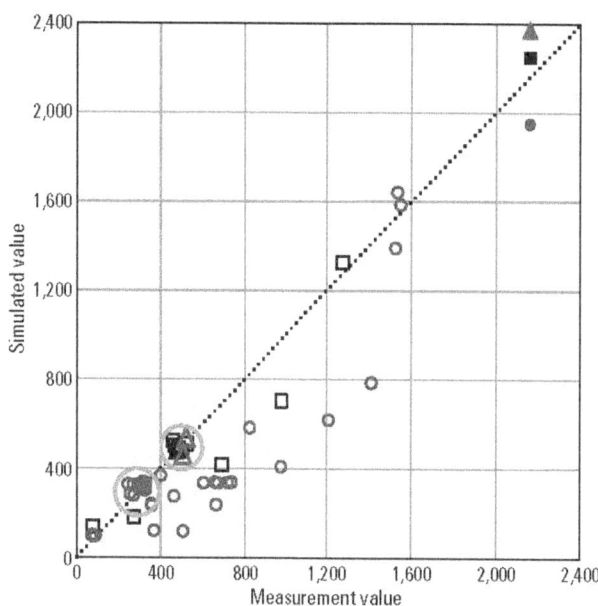

EXPLANATION

△ Pre-development water level

□ Early-time water level

○ Late-time water level

▲ Pre-development base flow

■ Early-time base flow

● Late-time base flow

······· Line of equal values

◯ Group 1 wells are within the circle

Units: Water levels in wells are in feet above North American Vertical Datum of 1988; and base flow is in gallons per minute.

Figure 27. Measured and simulated water levels in wells and base flow for the management scenario groundwater simulation model, Mosier, Oregon.

Commingling Well Vulnerability Maps

To understand where the aquifer system is vulnerable to commingling, the groundwater simulation model was used to compute change in basalt aquifer system storage resulting from leakage through commingling wells. Vulnerability maps were created by assuming that a larger loss in storage resulting from commingling equates to a more vulnerable area. Because final groundwater-levels in the aquifers are controlled by the elevation of the point where the water leaks out of the system and both aquifers slope upward toward the uplands, placement of a commingling well in a topographically low part of the system would likely result in lower final heads in many aquifers than placement of a similar well farther upslope (fig. 15B). As a result, the low land surface elevation areas of the aquifer system may be considered more vulnerable.

Vulnerability maps were constructed by computing the total change in storage in the basalt aquifers from pre-development conditions (no wells) to final steady-state conditions resulting from placement of a nonpumping commingling well in a single row and column of the model where more than one aquifer exists. The Grande Ronde flow top aquifer was excluded from this computation because few wells pump from it, and its hydraulic head is poorly defined over most of the study area. The rank, from largest to smallest, of the value of the change in storage was plotted at the center of each model cell (fig. 28). This is a single commingling well vulnerability map, and synergistic effects of commingling at multiple locations are not examined. The pattern of vulnerability shows that the area most vulnerable to commingling is generally coincident with the OWRD management area, which contains most of the Group 1 wells. The vulnerability map also indicates that the area to the west of the Maupin wrench fault may be vulnerable to commingling wells, although this area is currently largely undeveloped.

Simulation of Well Repair Options and Artificial Recharge/ Aquifer Storage and Recovery

To consider the potential value of repairing commingling wells, priority well repair zones were identified, and all wells in a particular zone were simulated as repaired. The modeling assumption that, unless a detailed interpretation of well construction information provides evidence to the contrary, each well passing through multiple aquifers commingles all aquifers intersected by the well, works well when predicting

aquifer response resulting from the net effect of commingling in many wells. However, there is no guarantee that any single well is actually commingling, because actual commingling depends on the local geology and the well construction method. Without knowledge about which wells are actually commingling, repair zones were defined (fig. 29) using the single well vulnerability map (fig. 28) and all wells in a defined region were simulated as repaired. This approach assumes that the exact location of commingling wells can be unknown, but the net effect of repairing any commingling wells in the zone still can be evaluated.

Two possible methods of recharging aquifers were considered: artificial recharge (AR) and aquifer storage and recovery (ASR) (State of Oregon, 1996). Artificial recharge is accomplished by applying water at the land surface and allowing it to infiltrate into aquifers. Aquifer storage and recovery is the process of injecting water through an injection well, and subsequently removing the water for use. Aquifer recharge from AR and ASR was simulated using the MODFLOW well package (McDonald and Harbaugh, 1984). Because the simulation method is the same for AR and ASR, the term AR/ASR is used in subsequent discussions of simulation results.

For evaluation of combinations of commingling well repair and aquifer recharge scenarios, the change in groundwater storage in the basalt aquifers was computed (assuming a simple confined storage coefficient), and expressed in terms of percent recovery (table 3). For example, if all groundwater levels recover to pre-development conditions, then it was assumed that the system was restored to 100 percent. The late time simulated conditions representing current pumping and commingling were assumed to be "restored" to 0 percent (not restored). The computed maximum amount of recovery due to repair of all commingling wells was 85 percent, where the remaining 15 percent is attributed to continued pumping. An uncertainty analysis was not done on the computed recovery, so the actual recovery due to repairs may vary from values summarized in table 3. However, the percent recovery may be used to evaluate the relative value of well repair and AR/ASR. The "Repairs only" column in table 3 summarizes the percent recovery of storage resulting from repairing the wells in high priority zones (fig. 29). The number of possibly commingling wells that is simulated as repaired for each scenario is listed in "Number of wells repaired" column (table 3).

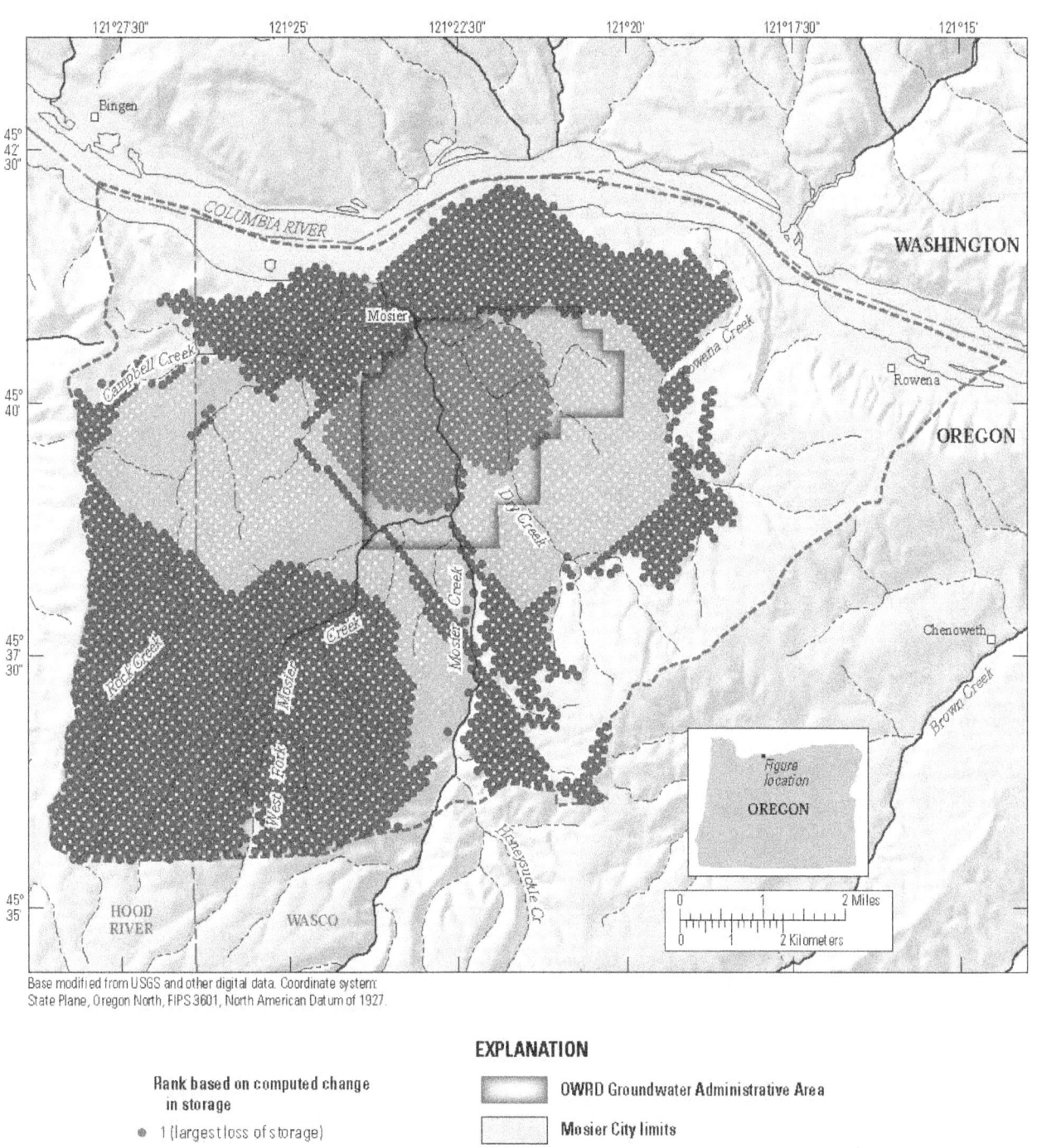

Base modified from USGS and other digital data. Coordinate system:
State Plane, Oregon North, FIPS 3601, North American Datum of 1927.

EXPLANATION

Rank based on computed change
in storage

● 1 (largest loss of storage)

● 2

● 3

● 4 (smallest loss of storage)

▨ OWRD Groundwater Administrative Area

▢ Mosier City limits

------ Geologic model and groundwater-flow model boundary

——— Perennial stream

—·—·— Ephemeral stream

Figure 28.   Relative vulnerability of the groundwater system to commingling wells in the Mosier, Oregon, study area.

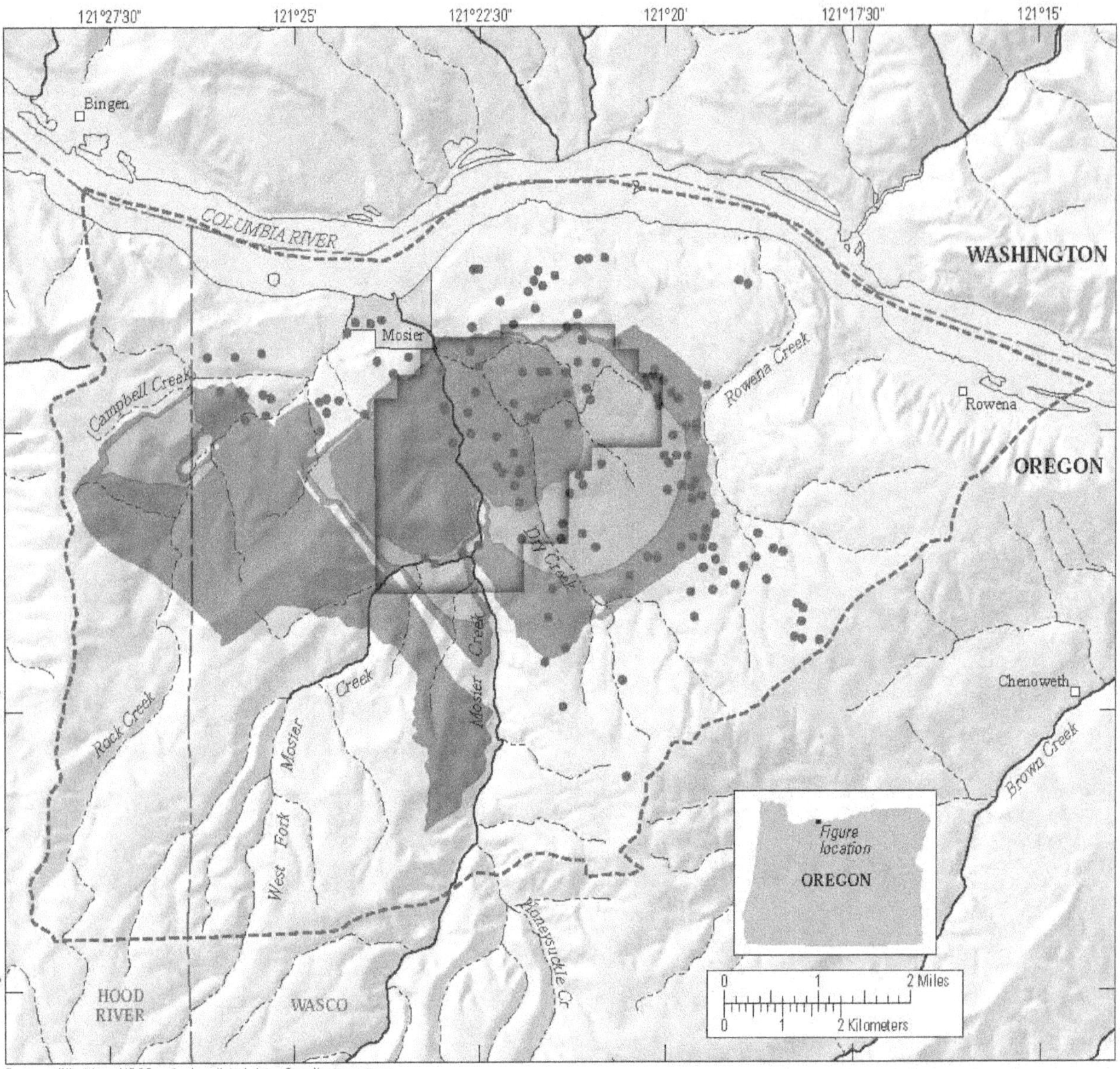

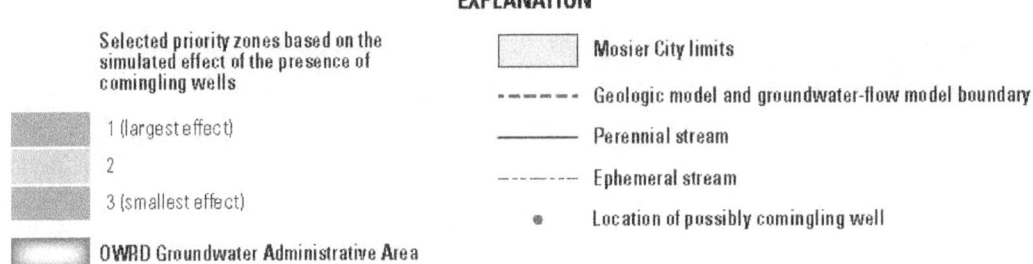

Figure 29.   Locations of wells currently simulated as commingled relative to ranked vulnerability zones in the Mosier, Oregon, study area.

The predicted recovery from AR/ASR was estimated by simulating recharge to each model cell representing an aquifer and computing the resulting change in aquifer system storage. Because the faults are strong flow barriers and the goal is to restore groundwater levels in the OWRD groundwater management area, AR/ASR was simulated into each CRBG aquifer model cell south of the Rocky Prairie thrust fault and east of the Maupin wrench fault (fig. 4). For all AR/ASR simulations, simulated injection rate was 250 gal/min for the 6 winter months, which corresponds to approximately 202 acre-ft of water each year or 21.5 percent of total annual pumping for the entire study area. Storage change resulting from AR/ASR in each model cell may be plotted for each basalt aquifer (for example, see fig. 30). Assuming AR/ASR would be at the optimal location, the maximum computed change in storage from all cells in all layers is reported in table 3 for comparison. The "Value of AR/ASR" column (table 3) shows the additional benefit of conducting AR/ASR in conjunction with some amount of well repair.

As commingling wells are repaired, the value of AR/ASR generally improves because less water is lost through intra-borehole flow. Additionally, the location of greatest change in storage varies depending on which wells have been repaired (fig. 30). Without well repair, injection in the OWRD administrative area yields little improvement in aquifer storage, but after repairing wells in the area, it becomes the best location to enhance aquifer-system storage using AR/ASR.

In all cases, simulations showed that application of AR/ASR to the Frenchman Springs aquifer would provide the most benefit to system storage overall owing to its large extent and the fact that commingling waters leaking from this unit will flow upward through shallower units. The maximum benefit location for the "No Repair" scenario is predicted to be near the crest of the Columbia Hills anticline (fig. 30A). Note, however, that the model boundary in this location prevents water leakage out of the model to the south. In reality, the overlying limb of the Chenoweth thrust fault to the south may allow water to leak out, indicating that the estimated 5.9 percent maximum recovery is likely an overestimation of the potential benefit of AR/ASR for the "No Repair" scenario.

Table 3. Summary of relative value of Artificial Recharge/Aquifer Storage and Recovery and commingling well repairs in the Mosier, Oregon, study area.

[Zones are shown in figure 29. Results should be used for comparison only. **Value of AR/ASR**: The difference between Repairs and Repairs plus AR/ASR columns. **Abbreviations**: AR, artificial recharge; ASR, aquifer storage and recovery]

Repair scenario	Percent recovery			Number of wells repaired
	Repairs only	Repairs plus AR/ASR	Value of AR/ASR	
All repaired	85.2	Not computed	Not computed	146
Zones 1, 2, and 3	54.2	63.1	8.9	82
Zones 1 and 2	23.1	30.4	7.3	50
Zone 1	11.1	17.2	6.1	25
No repairs	0	5.9	5.9	0

Evaluation of the Value of Repairs Targeting a Single Confining Unit

The final scenario considered was to repair the integrity of a single confining unit. In practice, a well open to three or more aquifers separated by confining units would not be repaired at a single confining unit, but the simulation results illustrate aquifer response that may be used when designing repair strategies. For example, in an area with multiple confining units, if one confining unit is penetrated by fewer wells than the other confining units, repair of this subset of wells effectively would hydraulically isolate the aquifers above and below the repaired confining unit, potentially providing substantial benefit. For the simulations, all commingling wells in repair zones 1, 2, and 3 were partially repaired. In each case, the repair separates two adjacent CRBG aquifers, but all other commingling still is assumed to exist.

Of the 82 wells simulated as commingling in these zones, at least one-half this number penetrates each confining unit (table 4). The percent recovery attributed to AR/ASR is fairly constant, but the percent recovery attributed to repairs starts out high for repair of the upper two layers, decreases for the third, but then increases again for the fourth. This irregular pattern can be explained with a simple conceptual model of storage.

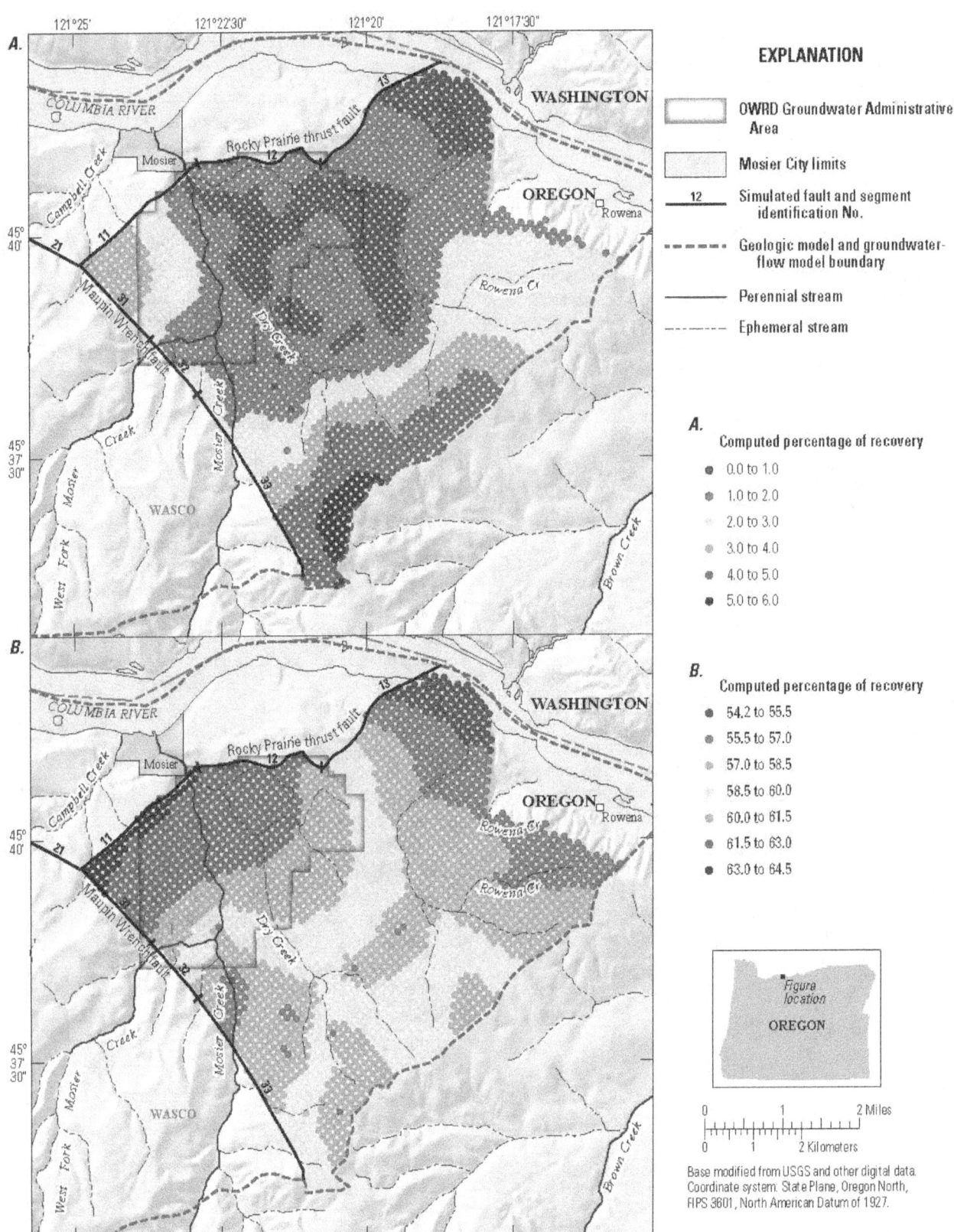

Figure 30.   Percentage of increase in aquifer-system storage resulting from Artificial Recharge/Aquifer Storage and Recovery into the Frenchman Springs aquifer at each location for repair scenarios (*A*) no repairs and (*B*) all commingling wells in zones 1, 2, and 3 repaired.

Table 4. Summary of simulated recovery resulting from targeted repair of a single confining unit in repair zones 1, 2, and 3 in the Mosier, Oregon study area.

[Results should be used for comparison only. **Value of AR/ASR:** The difference between the previous two columns. **Abbreviations**: AR, artificial recharge; ASR, aquifer storage and recovery]

Layers repaired	Percent recovery			Number of wells repaired
	Repairs only	Repairs plus AR/ASR	Value of AR/ASR	
Pomona flow interior confining unit only	29.9	35.1	5.2	69
Selah interbed only	34.0	40.6	6.6	72
Lolo flow interior confining unit only	8.4	15.0	6.6	57
Rosalia flow interior confining unit only	15.0	21.8	6.8	43
All four layers listed above	54.2	93.1	8.9	82

The single commingling well problem used to generate the vulnerability maps is conceptually illustrated by figure 31. Because water is added to each aquifer by recharge, if the commingling well did not exist, groundwater levels would rise until they reach the point where they naturally spill over (fig. 31A). The physical spill point for aquifers in the study area is where an aquifer intersects a spring or creek. Repair of any single confining unit allows water to fill any cross-connected aquifers until the lowest spill point is reached (fig. 31C through F). The amount of recovery depends on the geometry of the aquifer being filled.

In the Mosier aquifer system, there are many cross-connecting points (commingling wells), so complete repair of all wells in a zone (for example, fig. 29) will only provide partial recovery of the system (fig. 32). A more complex distribution of commingling wells (fig. 33) provides a simple explanation for the simulation results (table 4) that relies on the locations of commingling wells instead of the size and geometry of the aquifers.

In conclusion, repair of wells that raise the effective outfall for one or more aquifers will raise water levels in only those aquifers. Because the aquifer system is sloped, this indicates that sealing the system from the low end to the high end will provide a systematic improvement in water levels in all aquifers. Because the system currently is not in equilibrium, this strategy may also slow the rate of declines, because the gradient driving declines will be diminished.

Limitations of Management Option Analysis

A single set of model parameters was used for all management scenarios. This allowed a rapid comparative analysis of different management options, but the uncertainty in the magnitude of benefit has not been evaluated. Prior to adopting a particular management strategy, a predictive uncertainty analysis of the range of expected benefits could be performed.

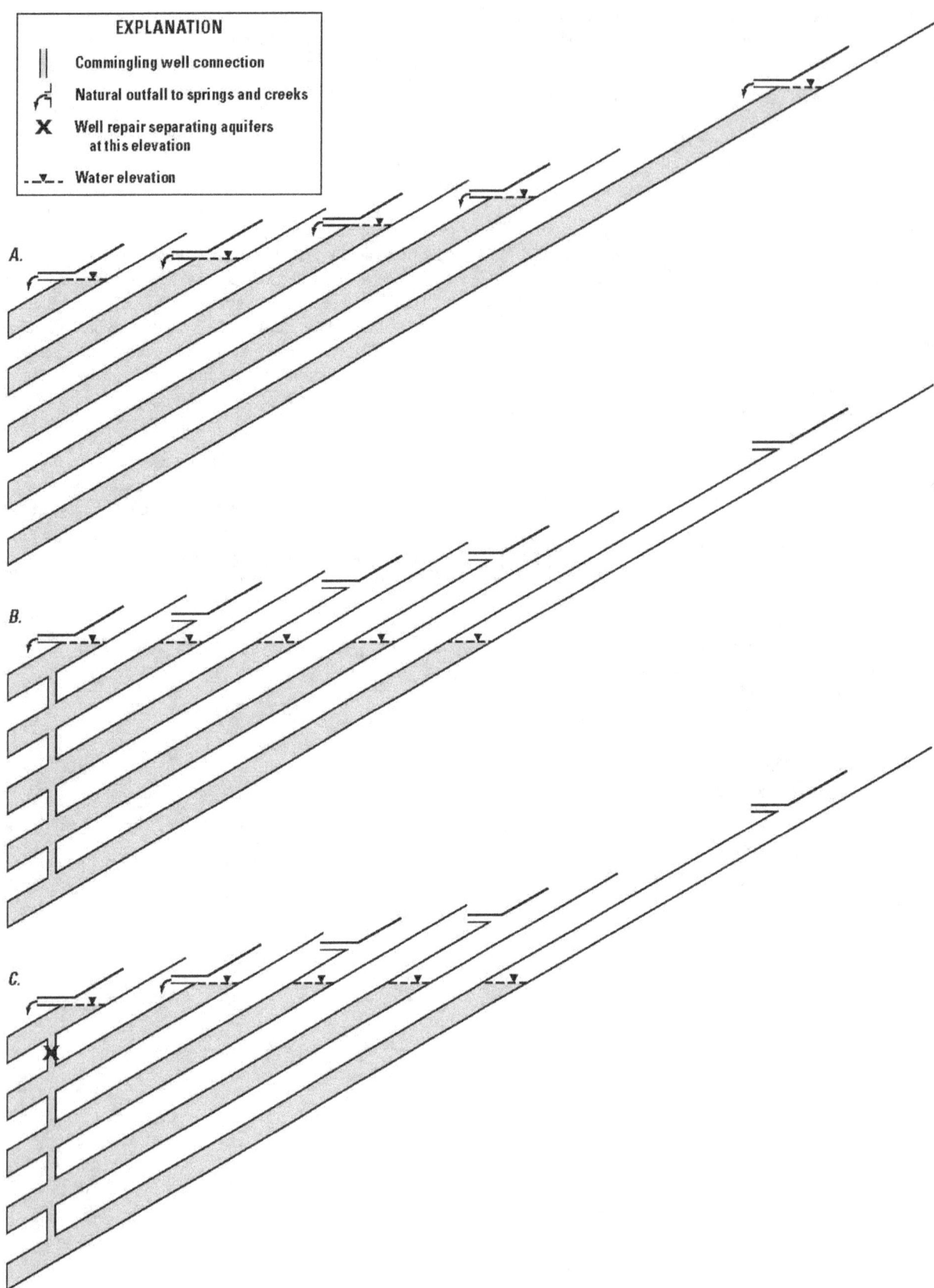

Figure 31.   A conceptual model of repairs to a single confining unit in the Mosier, Oregon, study area. The connecting pipe represents the commingling well effect. (*A*) No commingling well, (*B*) well commingling all aquifers, (*C–F*) repair of each confining unit.

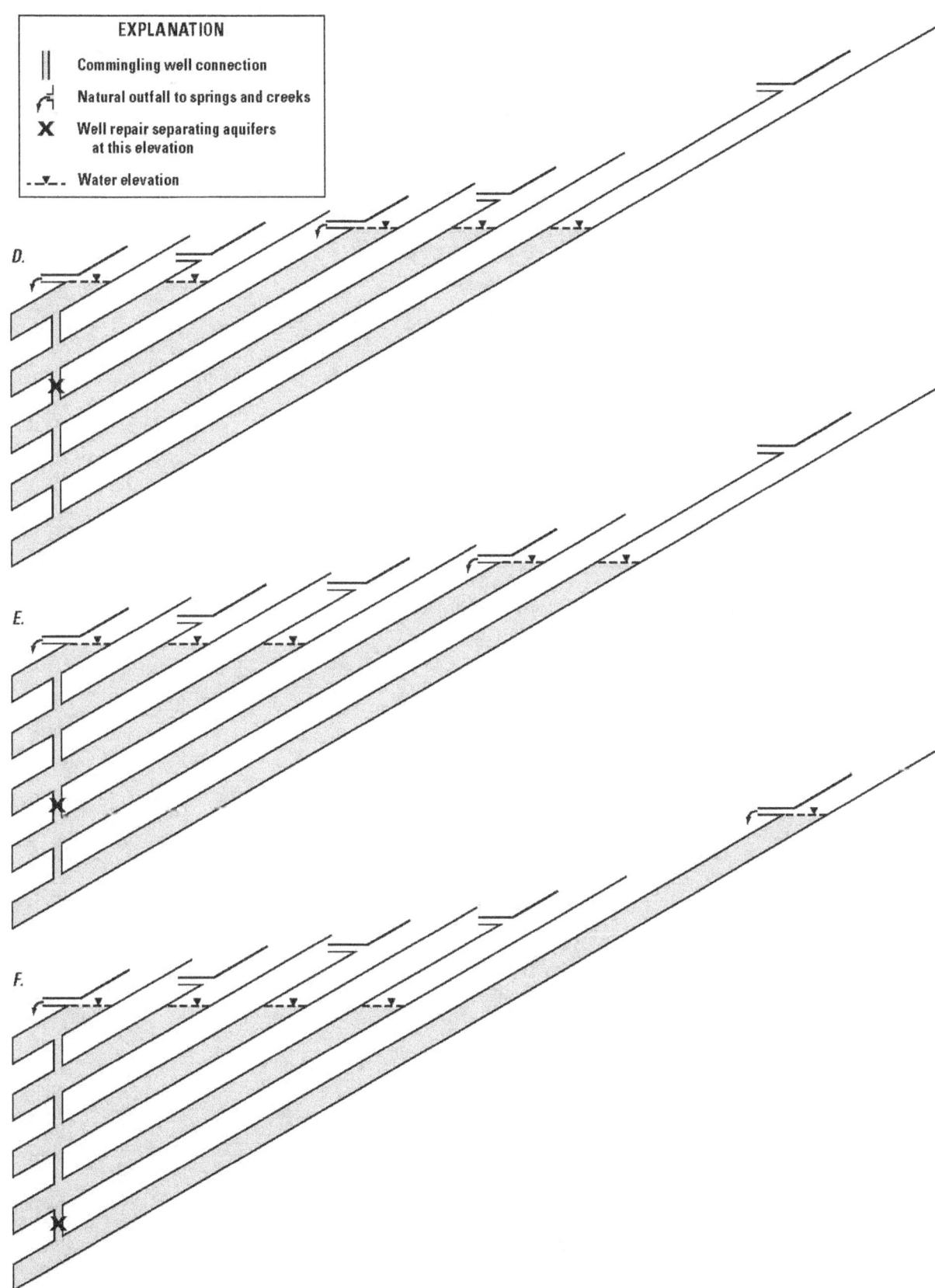

Figure 31.—Continued

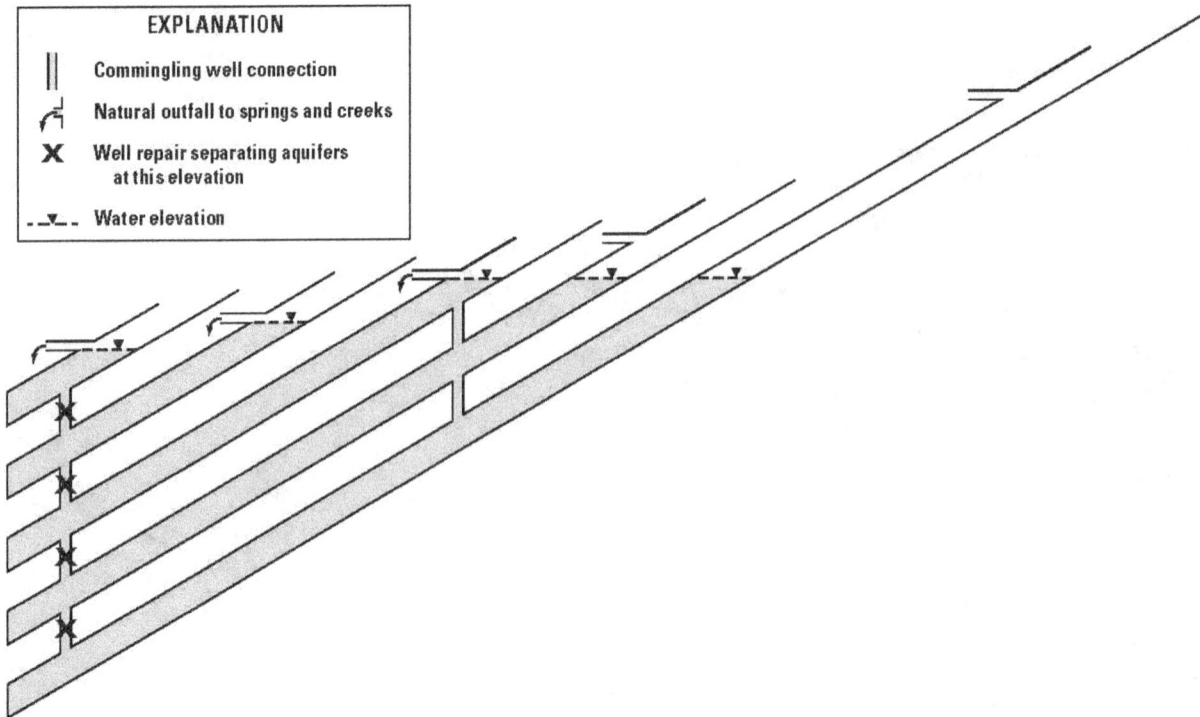

Figure 32.   Conceptual model of zonal repair of commingling wells in the Mosier, Oregon, study area.

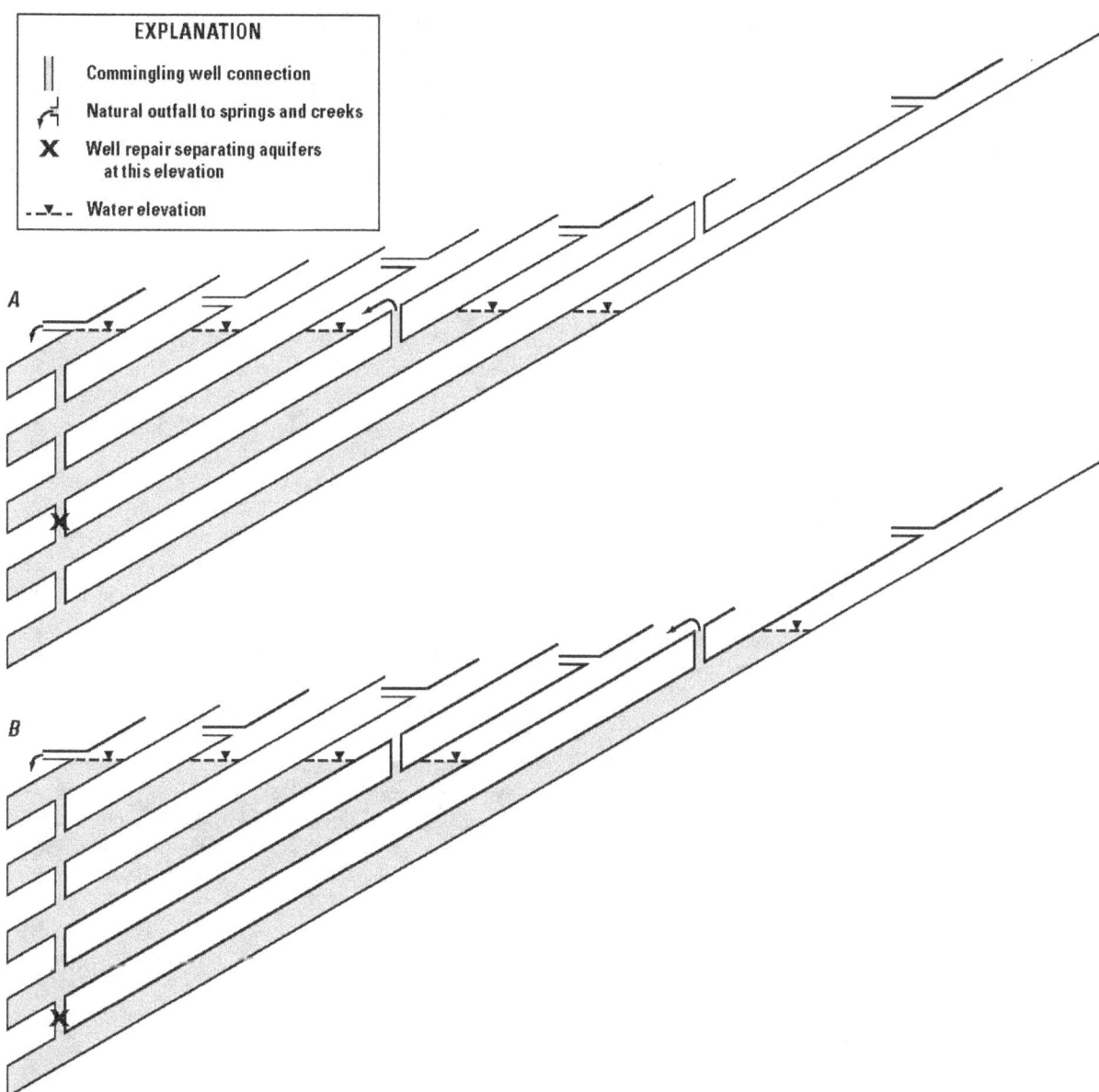

Figure 33. Conceptual model of how more complicated combinations of commingling wells and zonal repairs may explain the single confining-unit repair simulation results for the Mosier, Oregon, study area. (*A*) The relatively poor system recovery associated with repair of the Lolo confining unit (table 4) may be associated with commingling outside the zone of repair, and (*B*) a better system recovery associated with repair of the Rosalia confining unit (table 4) may be associated with a higher elevation controlling commingling outside the zone of repair.

Conclusions and Suggestions for Future Work

Groundwater simulations show that commingling wells are a significant and likely dominant cause of groundwater level declines in the study area. Further, the model allowed evaluation of vulnerability and possible management scenarios for restoration of aquifer water levels. The main conclusions are that the value of artificial recharge or aquifer storage and recovery will be greater if at least some commingling well repairs are accomplished first and that there are locations where repair of commingling wells is of higher value than in other areas. Highest vulnerability and highest value wells for repair are generally near the Oregon Water Resources Department administrative area with diminishing vulnerability and value upslope to the south.

Although the groundwater-flow simulation model was adequate for the analyses described in this report, the model has limitations that prevent its use for transient simulations, making it inadequate to predict the time necessary for the system to recover. The time for recovery likely will be different from the time it took for the system to decline because the rate of decline was controlled by how fast water leaks from the system, but if the system were completely repaired, the system recovery time will be controlled by the rate of recharge. In practice, the rate of recovery also will be limited by the rate of repair.

Potential groundwater-flow model improvements include better time-variable representation of commingling wells, the ability to represent the basalt aquifers as unconfined to allow dewatering of these aquifers to supply other aquifers through commingling wells, and representation of the apparent hydraulic conductivity contrast upslope of the Oregon Water Resources Department administrative area. Model calibration can be improved by establishing stream gages at the mouths of Mosier, Rock, and Rowena Creeks.

If commingling well repairs are accomplished, groundwater levels and streamflow also should be monitored. Ideally, a specialized monitoring network would be developed. These monitoring wells should be open to only one aquifer, and there should be a minimum of one monitoring well per aquifer. A potential alternative would be to install recording submersible pressure transducers at select wells in aquifers during well decommissioning. Collection of this data can become valuable for evaluation of the repair and recovery process.

Acknowledgments

Many individuals, groups, and agencies contributed to this work. Their contributions took many forms including time, resources, support, and guidance. The authors gratefully acknowledge all of these contributions and their importance to the successful completion of this work. A special acknowledgment is due the Mosier Watershed Council for their tireless pursuit of a community-supported plan that will ensure a sustainable supply of water for the Mosier Valley. The Council, a volunteer group of concerned citizens, has been led in this work by former chair Bryce Molesworth (Mr. February) and current co-chairpersons, Peter Kinsey (Mr. March) and Wade Root (Mr. January). The Council also had the support of Coordinators Jennifer Clark and Kate Conley from the Wasco County Soil and Water Conservation District (SWCD); their persistence and dedication toward developing support from State and Federal agencies has been the key to the success of the Council's work. We also acknowledge the unwavering support of Ron Graves, manager of the Wasco SWCD, who ensured that resources were available to the Council as they pursue a water management plan for the Mosier Valley.

The important contributions to the scientific understanding of the study area developed during the past 25 years by Ken Lite from the Oregon Water Resources Department are gratefully acknowledged. Conversations with Ken have been extremely helpful to the development of many of the ideas presented and approaches utilized in this work. Other scientists also helped by providing their data and insights from previous studies of the Mosier area. Most notably, these scientists include Gay Jervey, Rick Keinle, and Terry Tolan.

Other individuals who have provided valuable help and consultation include Tycho Granville (Wasco County Planning Department), Matthew Koerner (City of Mosier), John Grimm (John Grimm and Associates), and John Selker (Oregon State University).

Our final acknowledgement is to the orchardists and residents of the Mosier Valley who showed unflagging support for the development of a better scientific understanding of the water resources of the area. They supported the study in many ways, including allowing us access to their private wells for measurements and data collection, but also by attending meetings where they asked questions and provided direction on how to meet the needs of the community for information.

References Cited

Anderson, J.L., 1987, The structure and ages of deformation of a part of the southwest Columbia Plateau, Washington and Oregon: Los Angeles, University of Southern California, Ph.D. thesis, 272 p.

Bela, J.L., 1982, Geologic and neotectonic evaluation of north-central Oregon—The Dalles 1° × 2° Quadrangle: Oregon Department of Geology and Mineral Industries, Geological Map Series, GMS-27.

Burns, E.R., Morgan, D.S., Peavler, R.S., and Kahle, S.C., 2011, Three-dimensional model of the geologic framework for the Columbia Plateau Regional Aquifer System, Idaho, Oregon, and Washington: U.S. Geological Survey Scientific Investigations Report 2010-5246, 44 p.

Cleveland, W.S., Grosse, Eric, and Shyu, W.M., 1992, Local Regression Models, Chap. 8 of Chambers, J.M., and Hastie, T.J., eds., Statistical models in S: Pacific Grove, Calif., Wadsworth and Brooks/ Cole Advanced Books and Software.

Doherty, John, 2005, PEST—Model-independent parameter estimation, user manual (5th ed.): Watermark Numerical Computing, 336 p.

Doherty, John, 2010, Addendum to the PEST manual: Watermark Numerical Computing.

Grady, S.J., 1983, Ground-water resources in the Hood basin, Oregon: U.S. Geological Survey Water-Resources Investigations Report 81–1108, 68 p.

Halford, K.J., and Hanson, R.T., 2002, User guide for the drawdown- limited, multi-node well (MNW) package for the U.S. Geological Survey's modular three-dimensional finite-difference groundwater flow model, versions MODFLOW-96 and MODFLOW-2000: U.S. Geological Survey Open-File Report 02–293, 33 p.

Harbaugh, A.W., 1990, A computer program for calculating subregional water budgets using results from the U.S. Geological Survey modular three-dimensional groundwater flow model: U.S. Geological Survey Open-File Report 90–392, 46 p.

Harbaugh, A.W., Banta, E.R., Hill, M.C., and McDonald, M.G., 2000, MODFLOW-2000, the U.S. Geological Survey modular groundwater model—User guide to modularization concepts and the Groundwater Flow Process: U.S. Geological Survey Open-File Report 00–92, 121 p.

Hill, M.C., and Tiedeman, C.R., 2007, Effective groundwater model calibration—With analysis of data, sensitivities, predictions, and uncertainty: Hoboken, N.J., John Wiley and Sons.

Hsieh, P.A., and Freckleton, J.R., 1993, Documentation of a computer program to simulate horizontal-flow barriers using the U.S. Geological Survey modular three-dimensional finite-difference groundwater flow model: U.S. Geological Survey Open-File Report 92–477, 32 p.

Jervey, G.M. 1996. Transition lands study area groundwater evaluation, Wasco County, Oregon: Mosier, Oreg., Jervey Geological Consulting.

Kienle, C.F., 1995, Hydrogeologic investigation transition lands study area: Report prepared for Wasco County Planning and Economic Development Office, Wasco County, Oregon, Northwest Geological Services, Inc., 49 p. plus appendixes.

Leavesley, G.H., Lichty, R.W., Troutman, B.M., and Saindon, L.G., 1983, Precipitation-runoff modeling system— User's manual: U.S. Geological Survey Water-Resources Investigations Report 83–4238, 207 p.

Leavesley, G.H., Restrepo, P.J., Markstrom, S.L., Dixon, M., and Stannard, L.G., 1996, The modular modeling system (MMS)—User's manual: U.S. Geological Survey Open-File Report 96–151, 200 p.

Lite, K.E, and Grondin, G.H., 1988, Hydrogeology of the basalt aquifers near Mosier, Oregon—A groundwater resource assessment: Oregon Water Resources Department Groundwater Report No. 33, 68 p. plus appendixes.

McDonald, M.G., and Harbaugh, A.W., 1984, A modular three-dimensional finite-difference groundwater flow model: U.S. Geological Survey Open-File Report 83–875, 528 p.

Newcomb, R.C., 1961, Storage of ground water behind subsurface dams in the Columbia River basalt, Washington, Oregon, and Idaho: U.S. Geological Survey Professional Paper 383-A, 15 p.

Newcomb, R.C., 1963, Ground water in the Orchard Syncline, Wasco County, Oregon: Oregon Department of Geology and Mineral Industries, The Ore Bin, v. 25, no. 10, p. 133-138.

Newcomb, R.C., 1969, Effects of tectonic structure on the occurrence of groundwater in the basalt of the Columbia River Group of The Dalles area, Oregon and Washington: U.S. Geological Survey Professional Paper 383-C, 33 p.

Oregon Climate Service, 2009, Climate data—Hood River and The Dalles: Oregon State University database, accessed September 13, 2011, at http://www.ocs.orst.edu/.

Oregon Water Resources Department, 2006, Water Rights Information System (WRIS): State of Oregon website accessed 2006 at http://www.wrd.state.or.us/OWRD/WR/wris.shtml.

Piper, A.M., 1932, Geology and ground-water resources of the Dalles region, Oregon: U.S. Geological Survey Water Supply Paper 659-B, 189 p.

Powers, D.L., 1987, Boundary value problems (3rd ed.): Orlando, Fla., Harcourt Brace Javonovich, Publishers, 419 p.

PRISM Group, 2010, Precipitation (normals): PRISM Climate Group, Oregon State University, accessed September 13, 2011, at http://prism.oregonstate.edu/products/viewer. phtml?file=/pub/prism/us_30s/grids/ppt/Normals/us_ ppt_1971_2000.14.gz&yr=1971_2000&vartype=ppt&mont h=14&status=final.

Prudic, D.E., 1989, Documentation of a computer program to simulate stream-aquifer relations using a modular, finite-difference, groundwater flow model: U.S. Geological Survey Open-File Report 88-729, 113 p.

Rantz, S.E., 1982, Measurement and computation of streamflow, vol. 1—Measurement of stage and discharge: U.S. Geological Survey Water-Supply Paper 2175, 284 p.

Reidel, S.P., Johnson, V.G., and Spane, F.A., 2002, Natural gas storage in basalt aquifers of the Columbia Basin, Pacific Northwest USA—A guide to site characterization: Richland, Wash., Pacific Northwest National Laboratory, 277 p., accessed November 5, 2010, at http://www.pnl.gov/ main/publications/external/technical_reports/PNNL-13962. pdf.

Rutledge, A.T., 1998, Computer programs for describing the recession of groundwater discharge and for estimating mean groundwater recharge and discharge from streamflow data—Update: U.S. Geological Survey Water-Resources Investigations Report 98–4148, 43 p., accessed September 13, 2011, at http://pubs.usgs.gov/wri/wri984148.

State of Oregon, 1996, Division 350: Aquifer Storage and Recover (ASR) and Artificial Groundwater Recharge: Oregon Water Resources Department, accessed December 15, 2011, at http://arcweb.sos.state.or.us/pages/ rules/oars_600/oar_690/690_350.html.

Swanson, D.A., Anderson, J.L., Camp, V.E., Hooper, P.R., Taubeneck, W.H., and Wright, T.L., 1981, Reconnaissance geologic map of the Columbia River Basalt Group, northern Oregon and western Idaho: U.S. Geological Survey Open-File Report 81–797, 33 p.

Tolan, T.L., Martin, B.S., Reidel, S.P., Anderson, J.L., Lindsey, K.A., and Burt, Walter, 2009, An introduction to the stratigraphy, structural geology and hydrogeology of the Columbia River Flood-Basalt Province—A primer for the GSA Columbia River Basalt Group field trips, in O'Connor, J.E., Dorsey, R.J., and Madin, I.P., eds., Volcanoes to vineyards—Geologic field trips through the dynamic landscape of the Pacific Northwest: Geological Society of America Field Guide 15, p. 599–643, doi:10.1130/2009. fld015(28).

U.S. Census Bureau, 2004, Census 2000 TIGER/Line Data: U.S. Census Bureau database, accessed September 2008 at http://www.census.gov/.

U.S. Department of Agriculture, 2006, 2006 Oregon fruit tree inventory survey: U.S. Department of Agriculture, National Agriculture Statistics Service, Oregon Field Office report, accessed September 13, 2011, at http://www.nass.usda.gov/ Statistics_by_State/Oregon/Publications/Fruits_Nuts_and_ Berries/FTI-2006.pdf.

Viger, R.J., and Leavesley, G.H., 2007, The GIS Weasel user's manual: U.S. Geological Survey Techniques and Methods, book 6, chap. B4, 201 p.

Welty, J.R., Wicks, C.E., and Wilson, R.E., 1969, Fundamentals of momentum, heat, and mass transfer: New York, John Wiley and Sons, Inc., p. 697.

Appendix A. Development of a Three-Dimensional Hydrogeologic Framework Model

A three-dimensional hydrogeologic framework model was constructed using all available information. The foundation of the hydrogeologic framework model is a geologic framework model constructed from available surficial and structural geologic maps and geologic interpretations from 318 well logs (fig. 2) (Newcomb, 1969; Grady, 1983; Kienle, 1995; and Jervey, 1996). The following summarizes the technical aspects of the geologic modeling process and conversion of geologic model units to groundwater-flow model units (see Hydrogeologic Framework, figs. 3–5).

A.1—Motivation and Methods

The geologic framework modeling process was iterative (fig. A1), consisting of three interpretive steps: (1) creation of trend surfaces from compiled data, (2) building a 3-dimensional geologic model using the trend surfaces, and (3) analysis of the resulting model to evaluate how well the resulting geologic model matches the data. At the end of step 3, if the results are deemed inadequate, then the process is repeated with appropriate alterations to steps 1 and 2. Step 2 uses geologic and other physical principles, such as the geologic laws of superposition and original horizontality, to construct the distribution of geologic units from surfaces representing geologic unit tops and bottoms. More details of the motivation and methods used here are provided by Burns and others (2011).

Geologic unit tops were modeled as trend surfaces (fig. A1B), and it is assumed that the bottom of any unit is the top of the underlying unit. Trend surfaces were used for five reasons:

1. Well logs from which geologic picks were selected are of variable quality, with inherent inaccuracy in the estimates of ground-surface elevation and depth to contact;

2. Geologic picks from well logs from the previous studies were occasionally in conflict with each other, indicating that some picks are erroneous;

3. Geologic tops encountered in boreholes represent point samples of an undulating paleotopographic surface;

4. Mapped geologic contacts are smooth lines drawn across the current topography, providing artificial variability in estimated geologic unit top elevation; and

5. An understanding of the trend of the surface and local mismatch between data and the trend is deemed to be more informative about aquifer-system geometry than locally noisy fits acquired using exact interpolation of possibly erroneous data.

An example of using a trend to model a geologic surface is shown in the workflow of figure A1B, where mismatch between the trend surface and the data provides information about location and offset of a fault.

A.2—Geologic Model Assumptions and Implementation

Trend surface modeling was accomplished iteratively (fig. A1B); the first step was data compilation and confidence weighting. Because no data were obviously of lower quality, either initially or during the interpolation, all data received an equal weight. In addition to well data, the surficial geologic map compilation was also used. Where a geologic contact line represents the top of a unit at land surface, the line was sampled at a high frequency using points along the line. The elevation of the contact at each point was assigned the value from a digital elevation map of the land surface topography. Because the geologic map was constructed by drawing a smooth line across undulating terrain, the sampled geology also displayed this undulation. It was assumed that this elevation was correct on average, so points representing the local median were subsampled from the surficial geology points, providing a data reduction and estimates of the typical elevation of the contact.

The geologic map sample points were merged with the well data to provide sets of points representing each geologic model layer. Smooth trend surfaces were fit to data representing the top of each geologic unit (step 2 of fig. A1B). This was accomplished with a 2-dimensional local estimation regression method, LOESS, which was written in S for implementation in S-Plus (Cleveland and others, 1992). As implemented, LOESS performs a local linear fit to the data with an intended goal being a symmetrical distribution of the residuals around zero, where the residuals are defined as the difference between the measured value and the modeled surface. The LOESS algorithm has a variety of options, but all options were set to the default except for span and degree.

A.

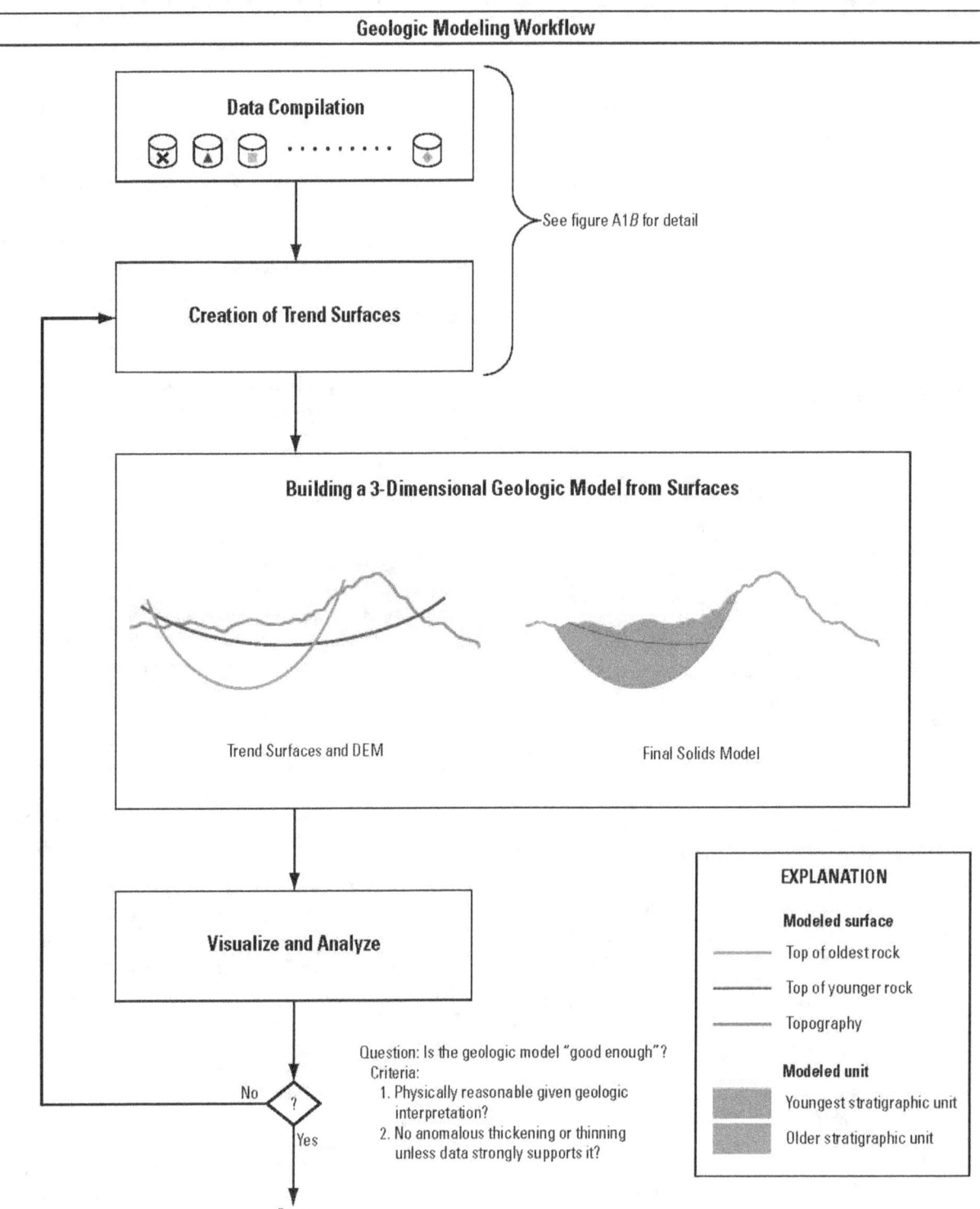

Figure A1.   The geologic modeling process. (*A*) Geologic modeling workflow, (*B*) details of trend surface modeling.

B.

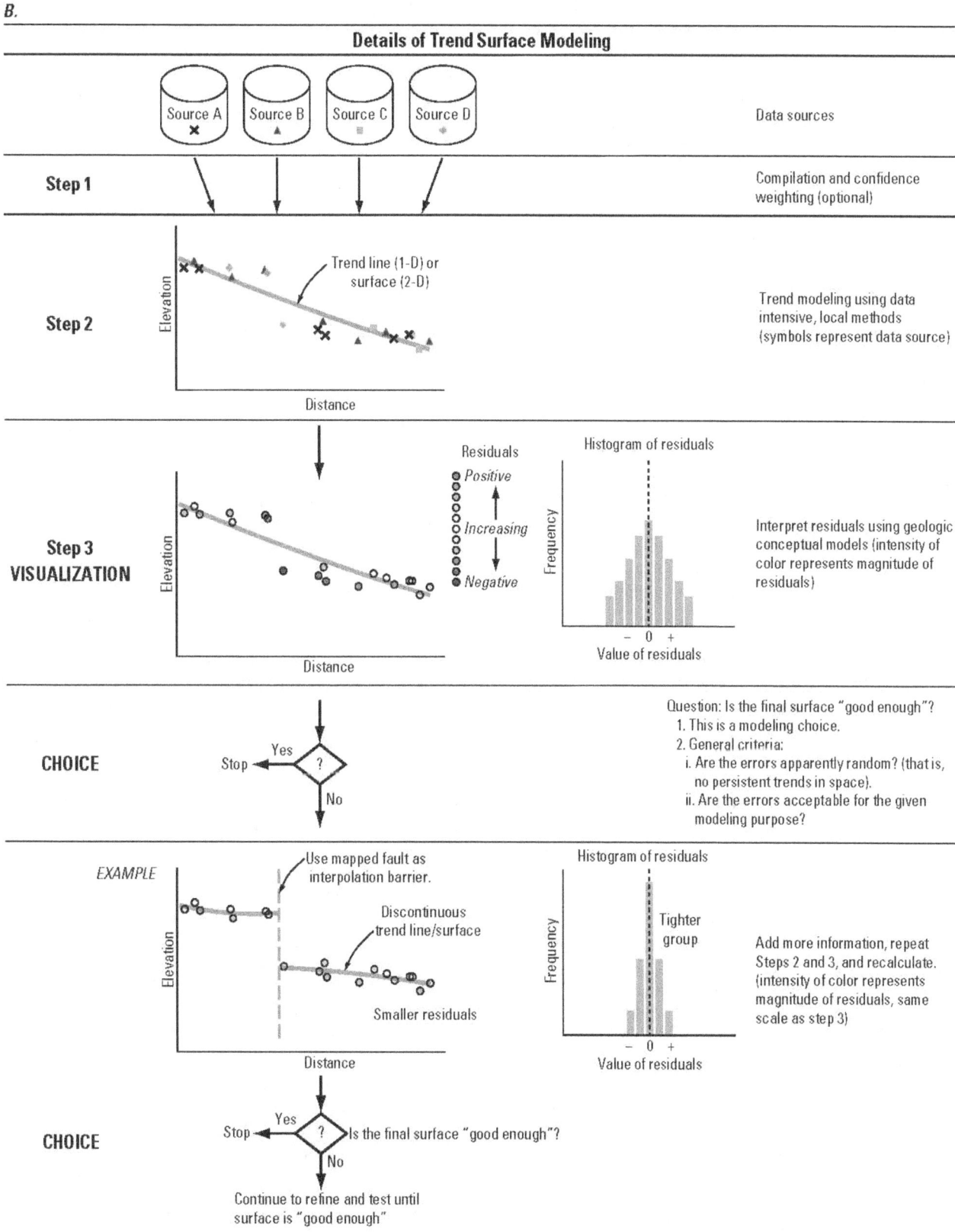

Details of Trend Surface Modeling

Source A ✕
Source B ▲
Source C ▦
Source D ◆

Data sources

Step 1

Compilation and confidence weighting (optional)

Step 2

Trend line (1-D) or surface (2-D)

Elevation

Distance

Trend modeling using data intensive, local methods (symbols represent data source)

Step 3 VISUALIZATION

Residuals
Positive
Increasing
Negative

Elevation

Distance

Histogram of residuals

Frequency

− 0 +
Value of residuals

Interpret residuals using geologic conceptual models (intensity of color represents magnitude of residuals)

CHOICE

Yes
Stop ← ?
No

Question: Is the final surface "good enough"?
1. This is a modeling choice.
2. General criteria:
   i. Are the errors apparently random? (that is, no persistent trends in space).
   ii. Are the errors acceptable for the given modeling purpose?

EXAMPLE

Use mapped fault as interpolation barrier.

Discontinuous trend line/surface

Elevation

Smaller residuals

Distance

Histogram of residuals

Frequency

Tighter group

− 0 +
Value of residuals

Add more information, repeat Steps 2 and 3, and recalculate. (intensity of color represents magnitude of residuals, same scale as step 3)

CHOICE

Yes
Stop ← ? Is the final surface "good enough"?
No

Continue to refine and test until surface is "good enough"

Figure A1.—Continued

Degree controls whether the local fitting function is linear or quadratic, and in all cases, linear was selected here. Span controls the percentage of data that goes into the interpolation at any given data point, and data are locally weighted during interpolation using a tri-cube weighting function with heavier weight nearer the interpolation point. As a result, increasing the span uses more data for the local estimation, resulting in a smoother trend surface. If the span is small, only data close to the interpolation point are used, and the surface becomes irregular and numerical edge effects can occur. An edge effect is where an incorrect trend supported by only a few measurement points is continued past the data, resulting in substantially incorrect estimates of the surface in that direction. The "best" span is defined here as the largest span for which the residuals appear to be randomly distributed in space (the smoothest surface with no strong trend in the residuals when they are plotted on a map).

Iterative trend surface modeling revealed three distinct fault-bounded geologic modeling blocks (fig. A2). The associated faults had significant offsets (greater than 200 ft) across known faults or fault groups (fig. 2). The boundaries of the geologic modeling blocks represent the approximate location of the large offsets of the Rocky Prairie thrust fault and the Maupin wrench fault(s). The Maupin wrench fault has high offset to the south, with little or no offset to the north. The geometry of this fault to the north is uncertain because of low data density on the west side of the fault. The well data were insufficient to resolve offset on other mapped faults, so only these two faults are represented in the geologic model.

Because the angle of plunge of the Mosier syncline through geologic modeling block 1 is not known and there are little data available to constrain the geologic model, inferred control points were added to the interpolation for this block to represent the syncline as a gently plunging feature. The resulting geologic model has high uncertainty in this block, so calculations and estimates for this side of the model implicitly have higher uncertainty. However, the modeled geometry retains the necessary hydraulic character to route recharge in a manner consistent with the best available geologic understanding of the watershed.

Sufficient data existed to generate surfaces representing the tops of the overburden, the Pomona Basalt, the Priest Rapids Basalt, the Frenchman Springs Basalt, and the Grand Ronde Basalt (figs. 2, 3, and A2, and appendix A.5). The Selah and Quincy-Squaw Creek interbeds were less frequently identified in well logs, indicating that they possibly were not present or missed during drilling. Well log interpretation of the Priest Rapids Basalt was inconsistent, with some studies identifying the Lolo and Rosalia sub-units, whereas others only used the more generic Priest Rapids designation. Because insufficient data were available for the interbeds and the top of the lower Priest Rapids Basalt, the tops of these units were modeled as proportions of the distance between the overlying and underlying tops.

The distance between the top of the Pomona Basalt and the top of the Priest Rapids Basalt is partially filled by the Selah Interbed, which is overlain by the Pomona Basalt. The interbed represents the soil and alluvium that was accumulating on top of the Priest Rapids flows until the Pomona was deposited over it. Because the Pomona basalt flowed in through the valley bottoms and over the lower ridges of the paleo-Mosier valley, the thickest parts of the Selah correspond to the valley bottom deposits with thinner or non-existent deposits preserved in the topographically higher areas. Data coverage of the Selah Interbed is sparse with the thickest deposits estimated to be approximately 170 ft, but more commonly between 30 and 50 ft in the OWRD groundwater administrative area (Lite and Grondin, 1988). To generate a physically reasonable layer to represent the hydraulic character of the Selah Interbed, the interbed was modeled as a fraction of the distance between the underlying Priest Rapids and overlying Pomona basalt tops. The fraction was scaled to zero feet of thickness in the uplands where the Pomona basalt pinches out and to 40 ft of thickness in the administrative area to provide a simple linear relation that matches the data reasonably well (fig. A3). In plan view (not shown) measured thickness is highly variable over short distances, with relatively thin deposits of this confining unit occurring not far from thick deposits. So even though a better fit might be achieved using more complicated relations, the simplified relation is likely adequate for use in the flow model. This is because there is much more uncertainty in hydraulic conductivity of the interbeds than in interbed thickness through which groundwater flows.

The thickness between the tops of the Frenchman Springs Basalts and the Priest Rapids Basalts is filled with two individual Priest Rapids Basalt flows and the Quincy-Squaw Creek interbed directly overlying the Frenchman Springs unit. Twenty-one wells had data on interbed thickness, and the sedimentary interbed had no strong or persistent spatial trend in thickness (fig. A4). Reported thicknesses ranged from 10 to 53 ft with an average value of 24.7 ft and a median value of 20 ft. It is assumed that presence of the Priest Rapids Basalt will preserve interbed deposits, so the interbed was modeled wherever Priest Rapids Basalt was present. If total distance between top of Frenchman Springs and Top of Priest rapids was greater than or equal to 30 ft, a constant thickness of 20 ft was modeled as interbed. Below 30 ft, interbed thickness is modeled as two-thirds of the total thickness (fig. A4).

Many wells contained picks for either or both of the upper and lower Priest Rapids flows, but when modeling the top of the upper Priest Rapids flow, it became apparent that a pick for top of the Priest Rapids unit occasionally was in fact a pick for the top of the lower Priest Rapids flow.

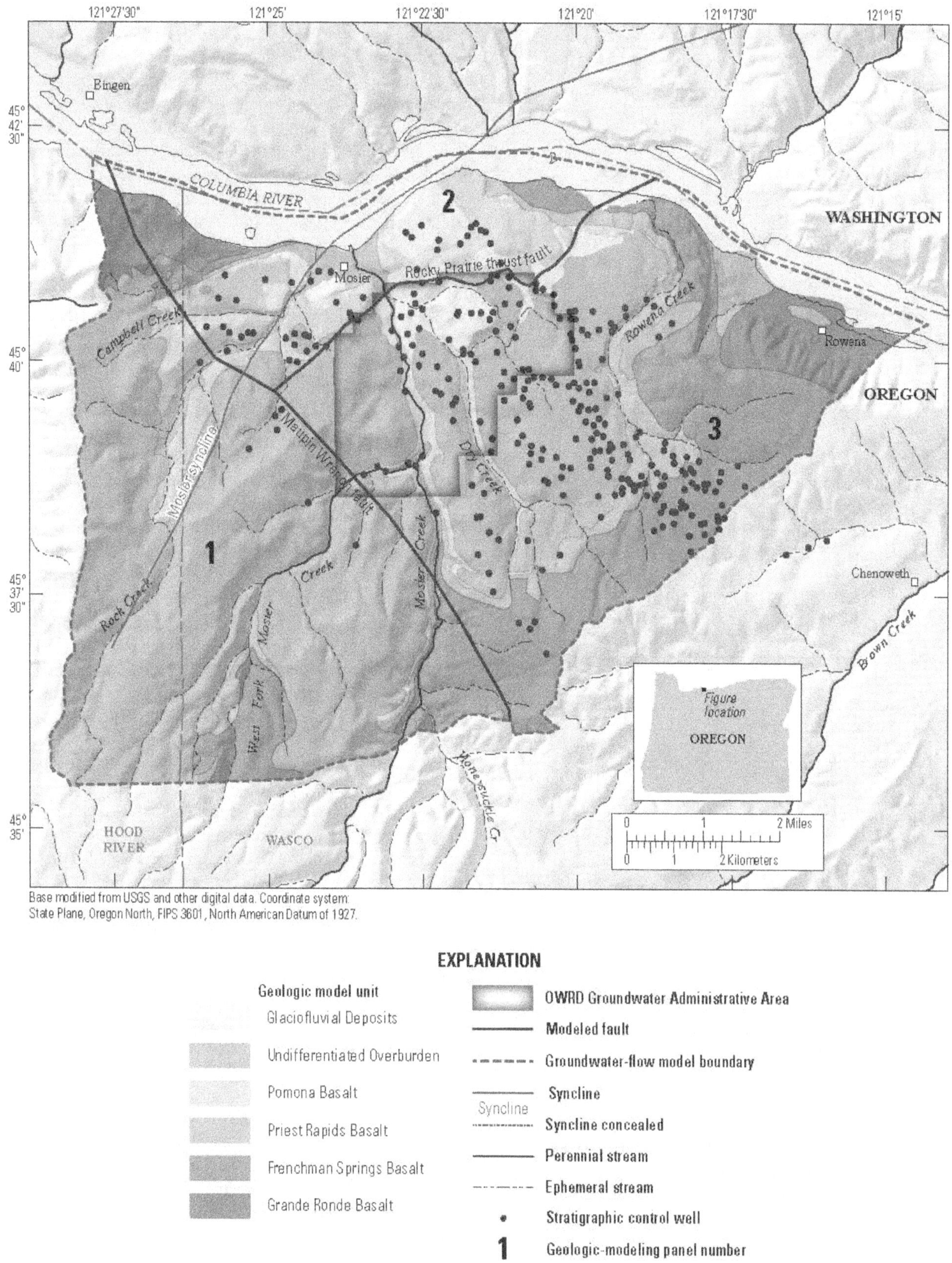

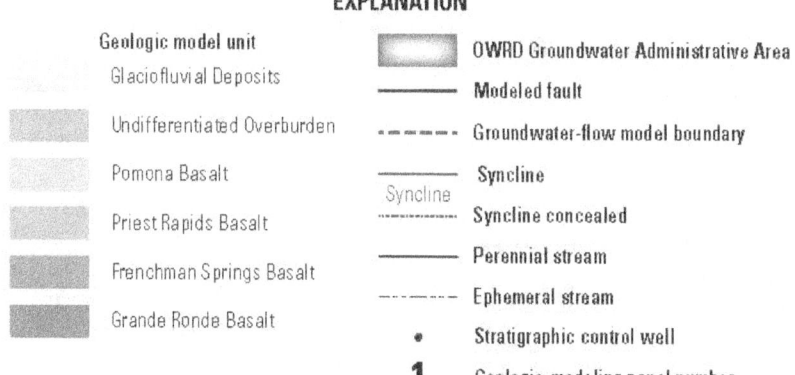

121°27'30"   121°25'   121°22'30"   121°20'   121°17'30"   121°15'

45°42'30"

COLUMBIA RIVER

Bingen

WASHINGTON

Mosier

Rocky Prairie thrust fault

Rowena Creek

2

45°40'

Campbell Creek

Rowena

OREGON

Mosier syncline

Maupin Wrap fault

3

Dry Creek

45°37'30"

Rock Creek

West Fork Mosier Creek

Mosier Creek

Chenoweth

Bown Creek

45°35'

1

Honeysuckle Cr.

HOOD RIVER   WASCO

Figure location

OREGON

0   1   2 Miles

0   1   2 Kilometers

Base modified from USGS and other digital data. Coordinate system:
State Plane, Oregon North, FIPS 3601, North American Datum of 1927.

EXPLANATION

Geologic model unit

Glaciofluvial Deposits

Undifferentiated Overburden

Pomona Basalt

Priest Rapids Basalt

Frenchman Springs Basalt

Grande Ronde Basalt

OWRD Groundwater Administrative Area

Modeled fault

Groundwater-flow model boundary

Syncline

Syncline concealed

Perennial stream

Ephemeral stream

•   Stratigraphic control well

1   Geologic-modeling panel number

Figure A2.   Simplified geology, wells with interpreted geology, and simulated faults corresponding to major structural features in the Mosier study area. Simulated faults were used to define separate geologic modeling blocks.

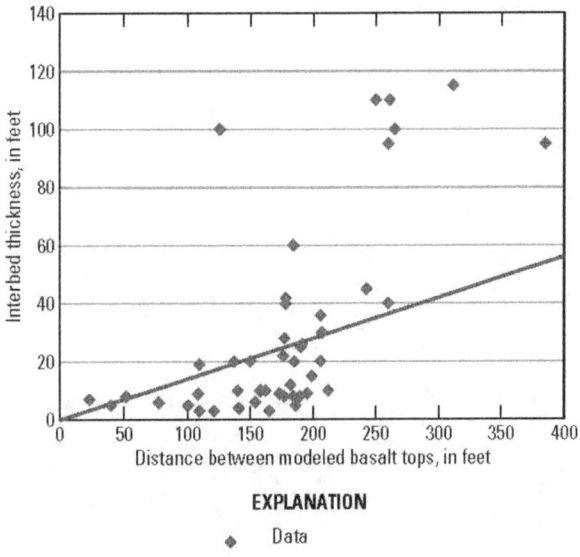

EXPLANATION

♦   Data

—— Modeled interbed thickness

Figure A3.   The relation between the thickness of the Selah interbed and separation between the top of the Pomona Basalt and the Priest Rapids Basalt. The line shows the simulated relation used in the geologic model.

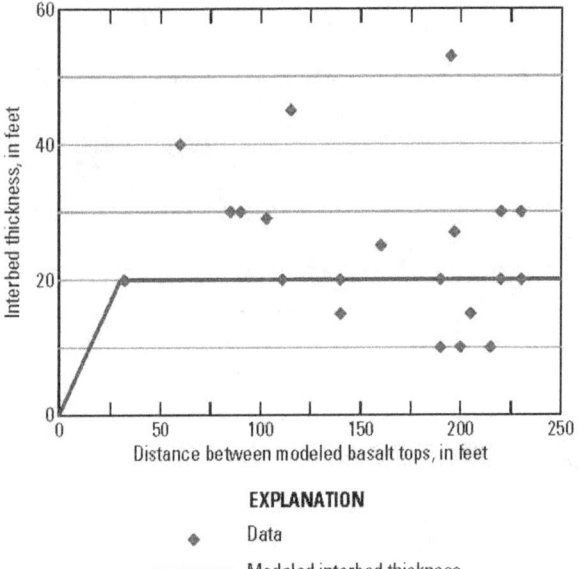

EXPLANATION

♦   Data

—— Modeled interbed thickness

Figure A4.   The relation between the thickness of the Quincy-Squaw Creek interbed and separation between the top of the Priest Rapids unit and Frenchman Springs unit. The line shows the modeled relation used in the geologic model.

To the maximum extent possible, these picks were corrected to provide a consistent set of data. To prevent inconsistencies with any erroneous points that may have been missed, the top of the Priest Rapids unit was modeled as a surface, and the demarcation between the two flows was modeled using thickness estimates from data where both flows were picked in the same wells (fig. A5). In these wells, the thicknesses of both the upper and lower units were examined, and the upper unit was approximately 70 ft thick with the lower unit exhibiting more variability. Additionally, if total thickness of the Priest Rapids flows plus interbed thickness was less than 82 ft, only one flow was typically present. For this reason, the demarcation between the units was computed by assuming that the lower unit and interbed constituted the entire thickness until the 82-ft threshold was exceeded, then the lower unit was modeled at a constant 82 ft thick until the upper unit was 70 ft thick. After the total value of 152 ft is reached, the upper unit is held at 70 ft of thickness with the balance being modeled as the lower unit. The lower unit is simulated as having a larger extent because the 82-ft threshold is larger than the 70 ft typical thickness of the upper unit and because it is convenient hydraulically to allow recharge to enter this unit directly for flow modeling. This precludes the need to rectify potential disconnects between focused recharge to the lower unit and the PRMS derived recharge, which inherently evenly distributes recharge across the PRMS hydrologic units. Contrarily, there is some preliminary geologic evidence that the upper unit can be of larger extent in some locations. The potential error introduced to the flow model is that more water is available from the lower unit at the expense of the upper unit, but because these units are separated by less than 70 vertical ft in the final flow model, and several wells draw water from multiple units, this error is likely negligible when considering the larger flow-system dynamics. The above relations are shown graphically compared to the data (fig. A5).

Only a few wells completely penetrate the Frenchman Springs Basalt and contact Grande Ronde Basalt. All of these wells are located in the upper part of the model area (generally, near the crest of the Columbia Hills anticline, fig. 2). As a result, the Frenchman Springs Basalt thickness and Grande Ronde unit top elevation are poorly defined except in this one small area. Even though the thickness is likely variable across the watershed, a typical upland value of 420 ft thickness of Frenchman Springs Basalt is used everywhere, defining the Grande Ronde unit top elevation everywhere. This simplified representation resulted in a corresponding poorer fit compared to other interpolated surfaces (table A1 and fig. A6). Further, there are no reliable estimates of uncertainty of this thickness because there are no Grande Ronde Basalt data in most parts of the watershed. However, the simplified formulation of both the Frenchman Springs and Grande Ronde Basalts is deemed adequate for groundwater-flow hypothesis testing subject to the conceptual model of the flow system.

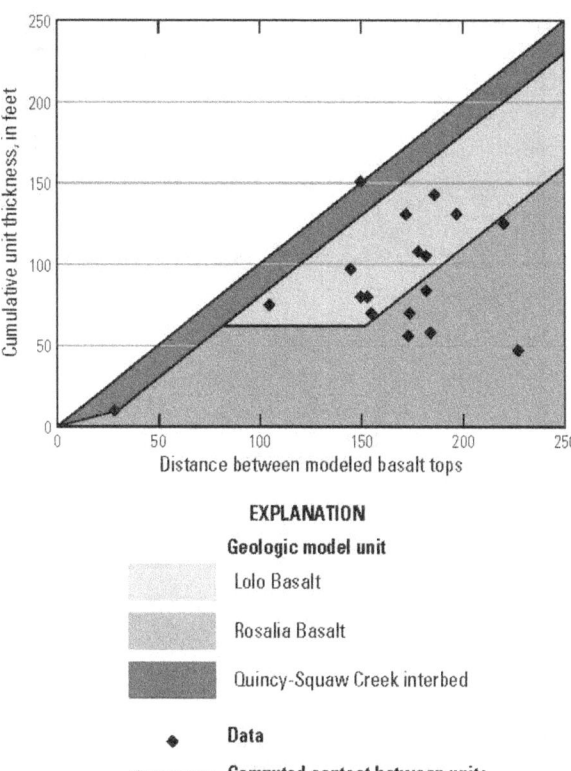

EXPLANATION

Geologic model unit

Lolo Basalt

Rosalia Basalt

Quincy-Squaw Creek interbed

♦ Data

—— Computed contact between units

Figure A5. The relations between the thickness of the lower Priest Rapids units (Rosalia) and the total thickness of Priest Rapids Basalt and the Quincy-Squaw Creek interbed. The lines show the rules used to construct the Lolo and Rosalia geologic model units as a function of the available thickness.

This conceptual model merely requires that recharge into Frenchman Springs and Grande Ronde Basalts be transmitted toward the Columbia River through the deep flow system.

The residuals of well data for the final modeled surfaces are generally symmetrically distributed (fig. A6) and are random in map view, with typical magnitudes less than 50 ft (table A1). Generally, inclusion of surficial geology points used in trend surface estimation will result in smaller computed variances, mean values closer to zero, and a more symmetrical distribution of residuals. However, the number of geologic points used in the interpolation is arbitrary because surficial geology can be sampled at an arbitrary interval. Because the number of sample points affects the computed statistics, only well points were used when computing and displaying summary statistics that may be used to infer model error.

An illustrative example of the role of geologic map sample points on geologic model fit results are shown for the Frenchman Springs flow top in geologic model block 3 (figs. A7 and A8). Figure A8C is data shown in A6D that is for geologic model block 3. The asymmetry of the histogram for the surficial geology points (fig. A8B) indicates that the top of the Frenchman Springs unit in outcrop is consistently higher than the trend of the well picks, indicating there may be some bias in selection of the top of the Frenchman Springs unit using drillers' logs.

Table A1. Summary statistics for well data residuals from the interpolation of each of the modeled basalt unit tops, Mosier, Oregon study area.

[**Variance** shown in units of square feet. All other quanities in units of feet. While including the surficial geology residuals would have a tendency to reduce variance and make the mean value closer to zero, the number of surficial geology points is a function of sampling methodology, so these data are not reported]

Unit top	Variance	Standard deviation	Number of points	Mean	Median	5th percentile	10th percentile	90th percentile	95th percentile
Pomona	2,404.0	49.0	165	13.8	10.6	-49.6	-36.2	78.9	106.5
Lolo	3,845.3	62.0	214	11.6	16.5	-107.9	-59.7	73.3	103.5
Rosalia	3,469.4	58.9	20	-12.7	-2.6	-96.8	-87.6	47.7	50.8
Frenchman Springs	3,986.9	63.1	240	-12.7	-3.8	-114.3	-79.8	51.9	65.6
Grande Ronde	11,589.7	107.7	39	-29.5	-37.2	-181.2	-162.6	119.6	124.9

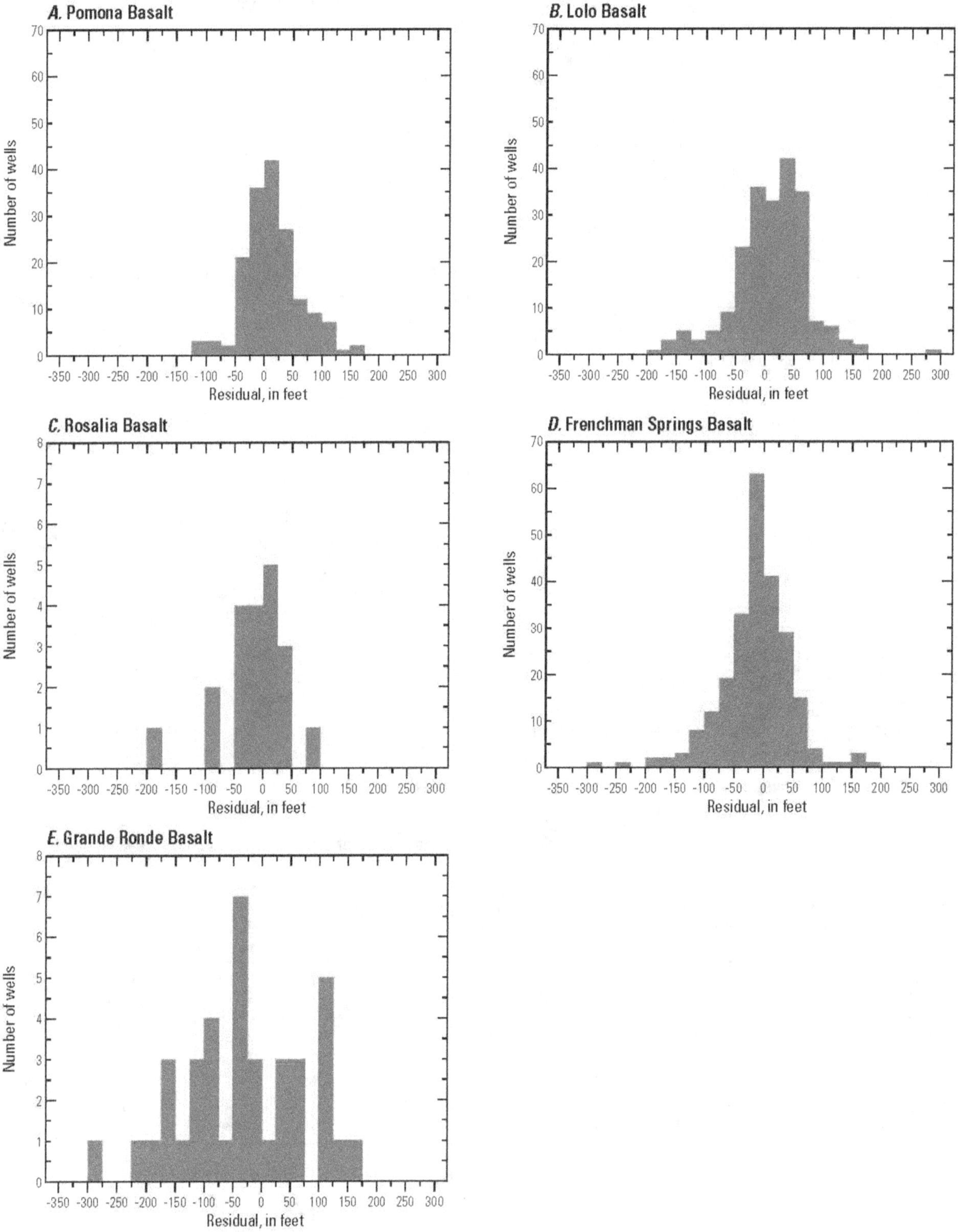

Figure A6.   Residuals for the wells only from the interpolation of each of the modeled basalt unit tops (*A*) Pomona Basalt, (*B*) Lolo Basalt, (*C*) Rosalia Basalt, (*D*) Frenchman Springs Basalt, (*E*) Grande Ronde Basalt.

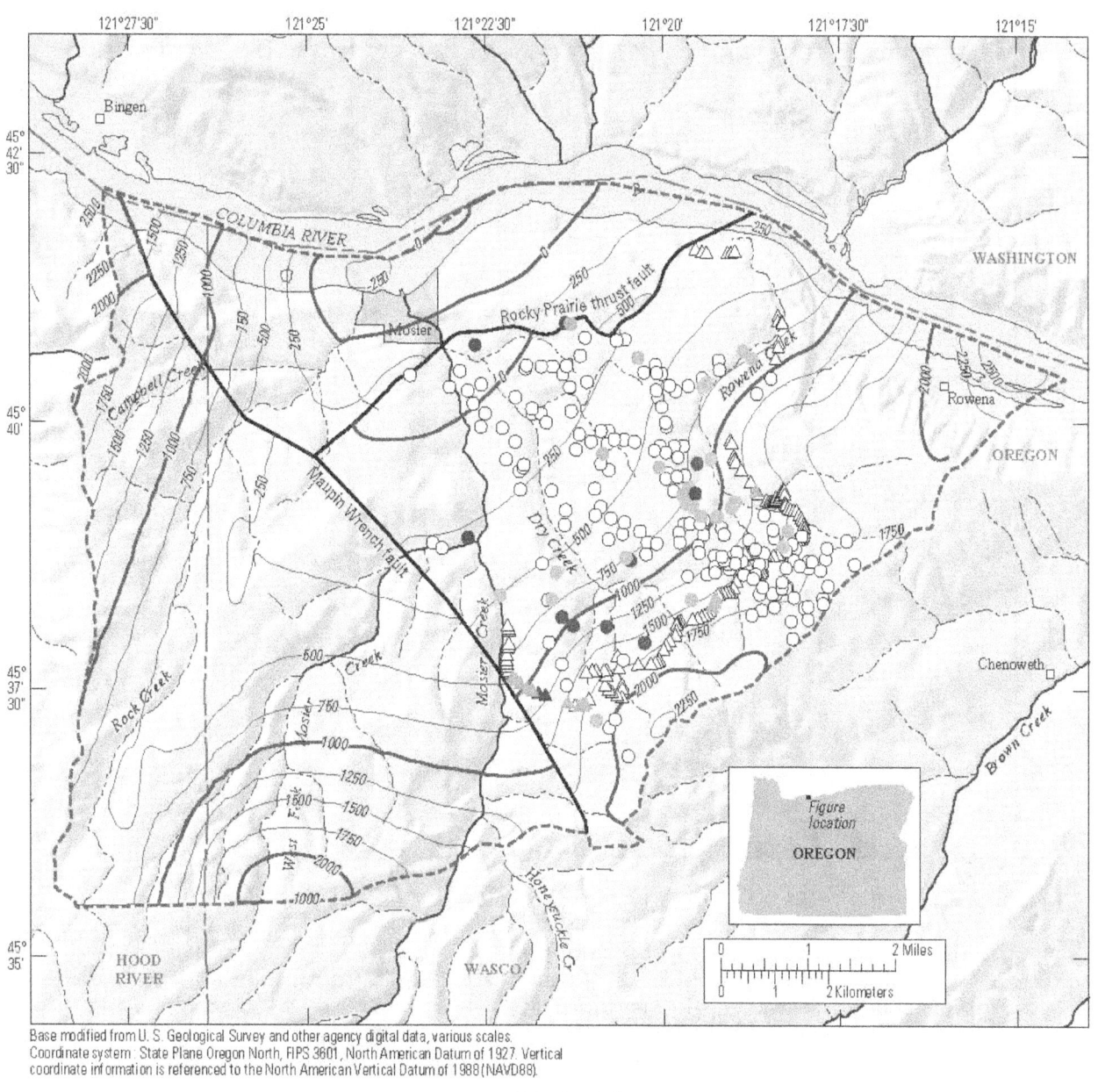

Base modified from U. S. Geological Survey and other agency digital data, various scales.
Coordinate system: State Plane Oregon North, FIPS 3601, North American Datum of 1927. Vertical
coordinate information is referenced to the North American Vertical Datum of 1988 (NAVD88).

EXPLANATION

Well data residual, in feet
- ● -309.2 to -150.0
- ◔ -149.9 to -75.0
- ○ -74.9 to 75.0
- ◕ 75.0 to 150.0
- ● 150.0 to 225.0

Geology map data residual, in feet
- △ -4.8 to 75.0
- ◭ 75.0 to 150.0
- ▲ 150.0 to 225.0

▭ Mosier City limits

—1000— Simulated trend surface elevation of top
of Frenchman Springs Basalt—Contour
interval 250 feet; bold line at 1,000-foot
intervals

- - - - - Groundwater-flow model boundary

——— Perennial stream

- - - - - - Ephemeral stream

——— Simulated fault

Figure A7. Distribution of residuals between both the geologic map and well data and the final trend surface for the top of the Frenchman Springs unit in geologic model block 3.

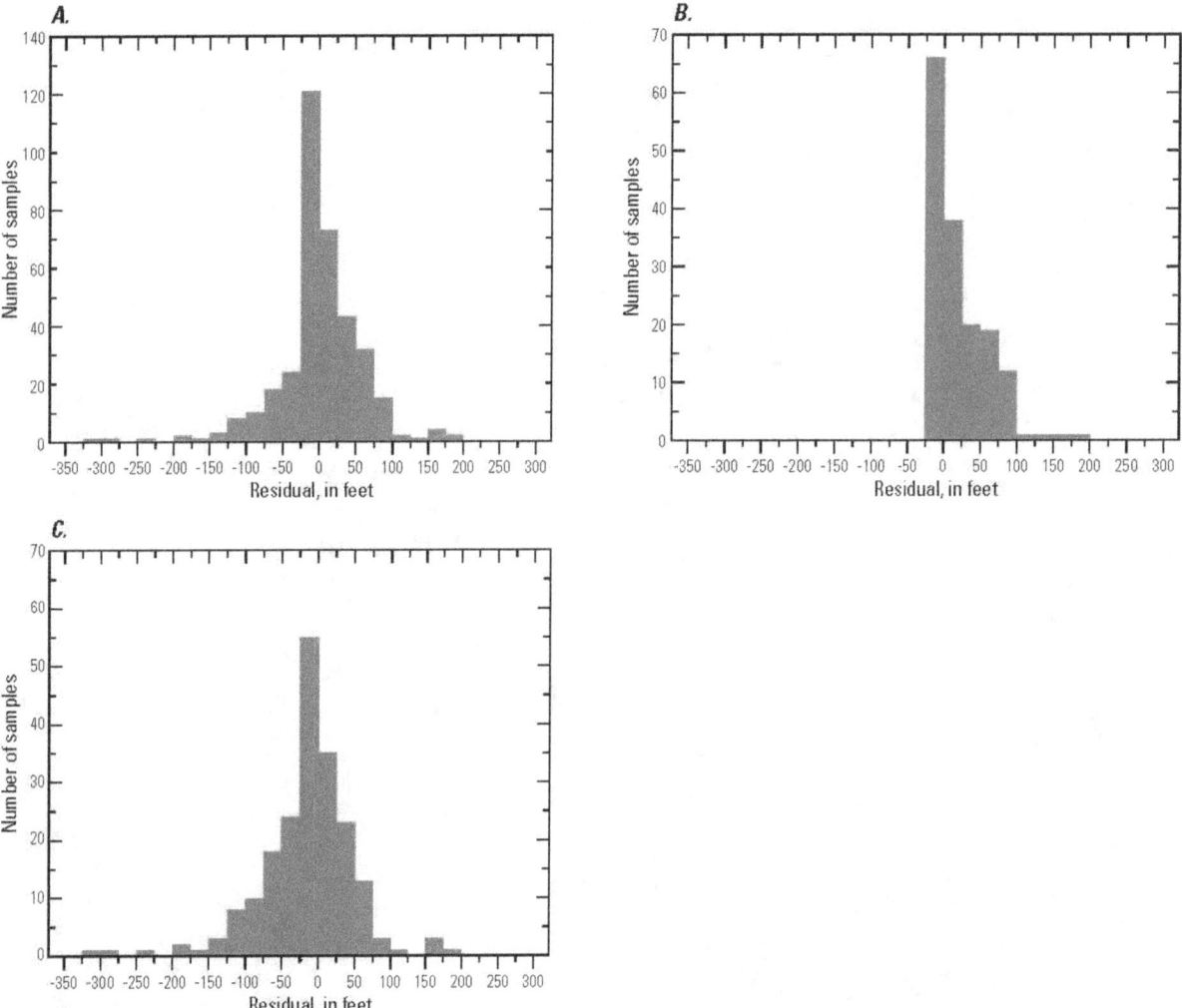

Figure A8.   Residuals between both the geologic map and well data and the final trend surface for the top of the Frenchman Springs unit in geologic model block 3. Histograms are for three data groups: (*A*) surficial geology sample points and wells, (*B*) surficial geology sample points only, and (*C*) wells only. The composite histogram shows that the modeling criteria is satisfied overall, but the skew in the other histograms indicates that the other data are potentially not consistent, indicating possible error in some well data.

Recall that the best trend surface is defined as the smoothest surface that removes most spatial trends in residuals and for which the distribution of residuals is symmetric with mean value of about zero. Rather than removing possibly erroneous data from the dataset, all data are used in the analysis. A color scheme was applied to figure A7 to highlight the effect of data outliers. Generally, spatial trends in these extremes are random, but some trends are discernible. Of particular interest is the cluster of highly negative residuals to the south. These negative values indicate that the smooth trend surface is higher than the data here. Except for these outliers, a smooth trend in the area adequately explains the observations, matching well data to the north and outcrop data to the south.

There are various possible reasons for these outliers. It is possible that the data represent a flow top other than the top of the Frenchman Springs unit. Recall that the Frenchman Springs Basalt in the area likely comprises four distinct flows. It is possible that the geologist that classified the top correctly identified the top of one of the lower Frenchman Springs units. In general, geologic picks with high negative residuals tend to occur in topographically lower areas, so if depth was used to identify the top of the unit (instead of elevation) then a lower Frenchman Springs flow top may have been identified as the top. Another possibility is that there is a fault or other geologic structure between the outliers and the remaining data, but there are no continuous mapped structures in the area (fig. 2). The last possibility considered here is that the data represent a local steep-sided bowl-shaped depression in the top of the Frenchman Springs unit. This violates the assumption that the surface is a fairly simple, smoothly varying surface. Because of the likelihood of erroneous interpretation of strata at some locations, the assumption of a simple smoothly varying surface is retained in the model for parsimony given the support of the remaining data. Retention of the high residual data in the fit statistics identifies data that can require further examination in the future.

Examination of fig A8C shows that the picks with outlier high negative residuals wells are generally 200–300 ft lower than the trend surface (fig. A8). In fact, these few outliers contribute a significant part of the computed variance of the residuals (table A2) and skew the histogram of the surficial geology picks (fig. A8B). Combination of the residuals from each data source, figures A8C and A8B, into a single composite data histogram (fig. A8A) shows that the fit criteria of symmetrical residuals with mean zero is generally achieved for the composite dataset (table A2).

Surficial geology data points vary smoothly in space, so inclusion of these residuals with well residuals for computation of fit statistics tends to reduce measures of spread (table A2), indicating that summary statistics based only on well data provide relatively conservative estimates of model error. A typical value of standard deviation for well residuals for each layer except Grande Ronde is about 60 ft (table A1). This measure of geologic model error generally applies in areas where there is some data support to ensure the trend surface is near the true value. In areas with poor data support, such as geologic-modeling panel 1, the estimated surface has greater uncertainty, and far from supporting data, should be viewed as reflecting the conceptual geologic model. Near structurally complex areas, close to faults with significant offset and tight folds, uncertainty also is greater due to the inability of a smoothly varying trend model to capture small-scale spatial variability. This is illustrated by the larger residuals near the Rocky Prairie thrust fault (fig. A7). When considering the geometry of the geologic units to be described in Mosier, a local error of 100 ft compared to the typical about 3,000 ft of relief of the surface to be described, corresponds to about 3 percent error. Because the median and mean errors are close to zero and there are no significant trends in residuals, the model-generated surfaces are correct on average.

Table A2. Summary statistics for residuals from the Frenchman Springs Basalt top interpolation in geologic-modeling panel number 3, Mosier, Oregon, study area.

[**Variance** shown in units of square feet. All other quantities in units of square feet]

	Variance	Standard deviation	Number of points	Mean	Median	5th percentile	10th percentile	90th percentile	95th percentile
Wells	4,269.5	65.3	203	-13.7	-3.3	-115.0	-87.3	51.1	62.8
Geology	1,261.2	35.5	159	23.3	0.9	-0.8	-0.3	75.3	87.8
Both	3,280.3	57.3	362	2.5	-0.0	-97.9	-62.7	60.1	81.1

A.3—Groundwater-Flow Model Units

The digital geologic model units were converted into groundwater-flow simulation model units (fig. 3) using estimates of the fraction of each unit occupied by hydrogeologic units. Each permeable basalt unit flow top is estimated as 10 percent of total flow thickness. Because the Frenchman Springs Basalt unit consists of four or five distinct flows in the area with insufficient data to allow delineation of these individual flows, the flow tops are modeled as one unit, consisting of the upper 10 percent of the total thickness. Because few wells penetrate the upper part of the Grande Ronde Basalt in the study area, only the flow top is modeled and the flow interior is treated as a no flow boundary, precluding the need to model it explicitly in the geologic model. Flow top thicknesses of 10–20 ft are not uncommon for sheet flows, and because the Grande Ronde is assumed to be a sheet flow in the study area, the Grande Ronde flow top is assumed to be a uniform thickness of 20 ft.

In this study area, the only basalt flow with a documented laterally extensive, permeable flow bottom is the Pomona Basalt (Lite and Grondin, 1988). The permeable part of the flow bottom has a much smaller footprint than the flow itself. It is generally coincident with the thicker parts of the Selah Interbed and is postulated to have formed when the hot basalt flowed across the wet valley bottom deposits. The footprint of the permeable zone in geologic model block 3 has an estimated areal extent between 4 and 6 mi^2 immediately south of the Rocky Prairie thrust fault with a maximum thickness of about 20 ft (Lite and Grondin, 1988). Further, prior to development, this zone was hydraulically isolated from the underlying Priest Rapids flow top aquifer. Because the Pomona Basalt flow was thickest in valley bottoms, the thickness of this unit was used to estimate a reasonable distribution of permeable flow bottom. Whenever modeled Pomona thickness exceeds 155 ft, permeable flow bottom is assumed to exist, and the excess thickness is scaled linearly so that the maximum thickness of the flow bottom is 20 ft. The resulting modeled aquifer has several nice properties: (1) it is about 4 mi^2 in the area generally identified by Lite and Grondin (1988), and about 8 mi^2 over the entire area; and (2) the footprint is completely contained within the Selah Interbed footprint, ensuring the Selah can act as an confining unit as supported by early data. The modeled aquifer overly simplified, and exact geometry may be somewhat different, but supporting data indicate the geometry is a reasonable representation for use in the hydrogeologic flow simulation model.

In an idealized geologic model, flow tops would be modeled as 10 percent of the thickness before truncating the model with surficial topography. This would ensure that the flow top intersected erosional stream cuts at the appropriate elevation. This would result in the flow interior being exposed in a thin band along the stream-cut wall, much like is observed in reality. Instead, based on flow modeling considerations, this was not done here. The flow top is modeled here as 10 percent of the final computed thickness at all locations, even at stream cuts and other erosional features that may occur on topographic highs. This provides many properties that aid in stability of the numerical flow model, allowing for robust estimation of parameters during automated sensitivity and predictive runs, and that are consistent with physical and other modeling assumptions. The properties aiding stability include:

1. Because PRMS provides a recharge estimate that is uniformly distributed across a model hydrologic response unit, preferential recharge pathways into system aquifers are not represented in the recharge field. Rather than identifying focused recharge areas and redistributing PRMS estimated recharge to these zones, simulating the upper 10 percent of every basalt unit as aquifer allows the distributed recharge to enter the aquifers as distributed recharge. This prevents high model heads from occurring when flux into confining unit model cells is prescribed, preventing the need to alter the hydraulic conductivity of these cells to achieve physically reasonable simulation results.

2. Connectivity between basalts and streams and drains is highest where the flow top is present in reality, and diminishes as basalts thin erosionally, providing a physically reasonable distribution of stream connection. This is especially true for the lumped Frenchman Springs Basalts, which is commonly about 400 ft thick consisting of multiple flows, but is modeled as a single flow top over a thick low permeability interior.

The overburden is laterally zoned into undifferentiated overburden and glaciofluvial deposits (fig. A2). Each zone is assumed to occupy the upper two groundwater-flow model layers (fig. 3). The Chenoweth Formation, which constitutes most of the undifferentiated overburden, is documented as generally having low permeability, as is expected of poorly sorted mud and ash deposits of volcanic origin. However, the base of this unit is occasionally coarse-grained and productive. Considering the morphology of debris flows, a gross oversimplification would be to conceptualize that the coarse, heavy deposits were funneled into the lowest path and to fell out much more rapidly than the finer deposits.

This gives a conceptual model of deposits that grade from fine to coarse from top down, and thicker sequences of coarse deposits in valley bottoms which also have thicker deposits overall. Accordingly, the overburden was divided into an upper 90 percent and lower 10 percent to allow for a relatively higher permeability base. Glaciofluvial deposits also retain the 90 to 10 percent split of groundwater-flow model layers, but because no data indicated that the upper and lower parts had dissimilar hydraulic properties, both groundwater-flow model layers were assigned the same hydraulic properties.

A.4—Groundwater-Flow Simulation Model Surfaces

Tops were computed for each of the groundwater-flow model units (fig. 3) at a 500 ft MODFLOW grid spacing (figs. A9–A22). The Grande Ronde aquifer unit is assumed to be 20 ft thick, defining the bottom of the flow model domain (20 ft below the surface shown in fig. A22). Each figure shows the extent of the MODFLOW model grid with cell color reflecting elevation and relevant boundary conditions displayed in appropriate cells. The set of lines from which the horizontal flow barriers are derived also is shown (figs. A11–A22).

For groundwater-flow simulation, model layers 3–11 have thin pseudo-cells that transmit water vertically between layers where hydrogeologic units have pinched out. These pseudo-cells are not shown on figures A11–A19, although they are a part of the active domain for groundwater-flow simulation. The reason for use of these cells is summarized in appendix E.3.

Upper overburden units (layer 1)

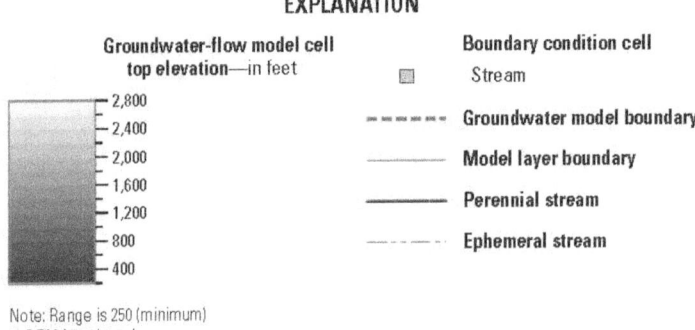

121°27'30"
121°25'
121°22'30"
121°20'
121°17'30"
121°15'
45° 42' 30"
45° 40'
45° 37' 30"
45° 35'
Bingen
COLUMBIA RIVER
WASHINGTON
Mosier
Campbell Creek
Rowena Creek
Rowena
OREGON
Dry Creek
Mosier Creek
Rock Creek
Mosier Creek
West Fork
Chenoweth
Brown Creek
Honeysuckle Cr.
Figure location
OREGON
0 1 2 Miles
0 1 2 Kilometers
HOOD RIVER
WASCO
Base modified from USGS and other digital data. Coordinate system: State Plane, Oregon North, FIPS 3601, North American Datum of 1927. Vertical coordinate information is referenced to the North American Vertical Datum of 1988 (NAVD88).
EXPLANATION
Groundwater-flow model cell top elevation—in feet
2,800
2,400
2,000
1,600
1,200
800
400
Note: Range is 250 (minimum) to 2,709 (maximum).
Boundary condition cell
Stream
Groundwater model boundary
Model layer boundary
Perennial stream
Ephemeral stream

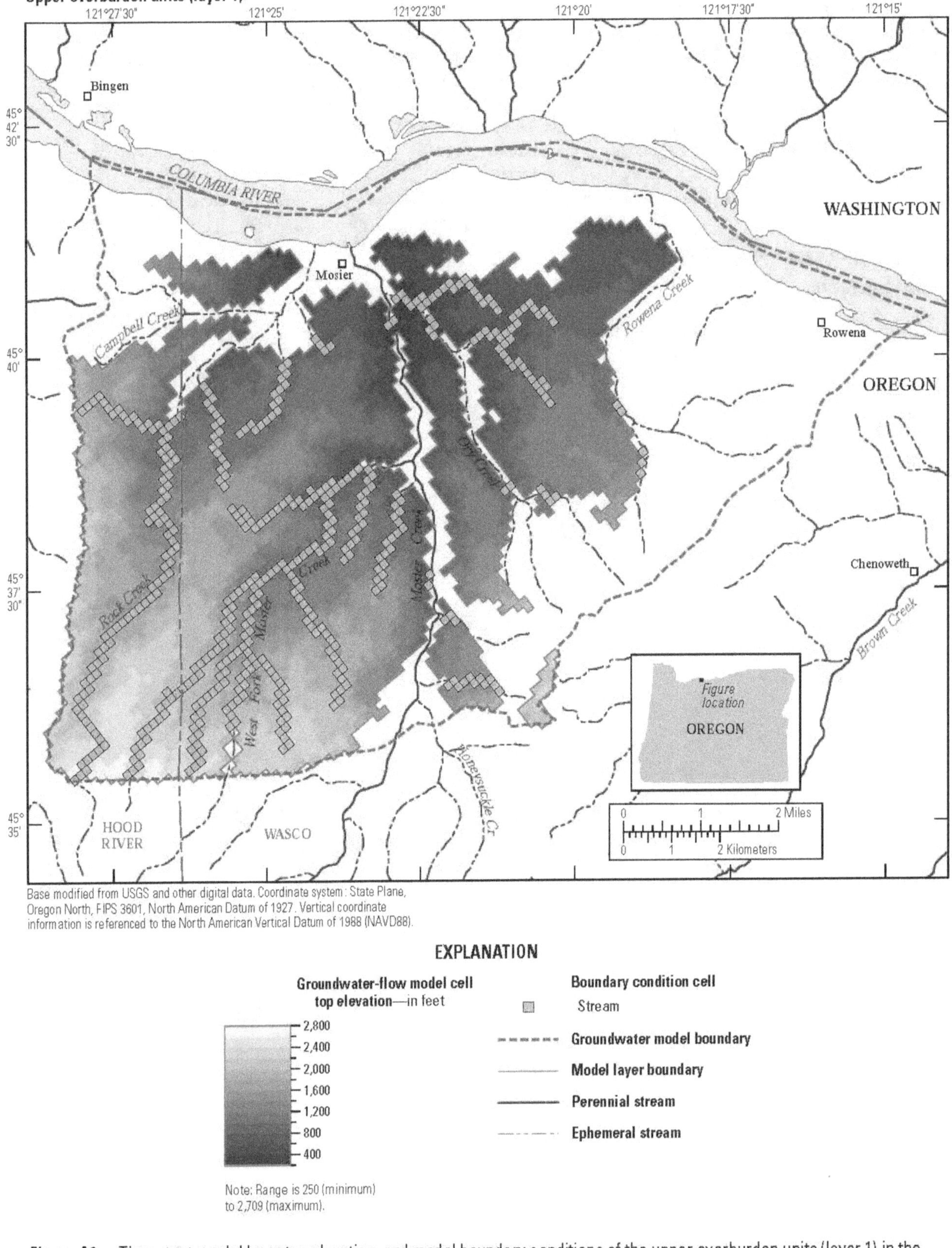

Figure A9. The extent, model layer top elevation, and model boundary conditions of the upper overburden units (layer 1) in the Mosier, Oregon, groundwater-simulation model area.

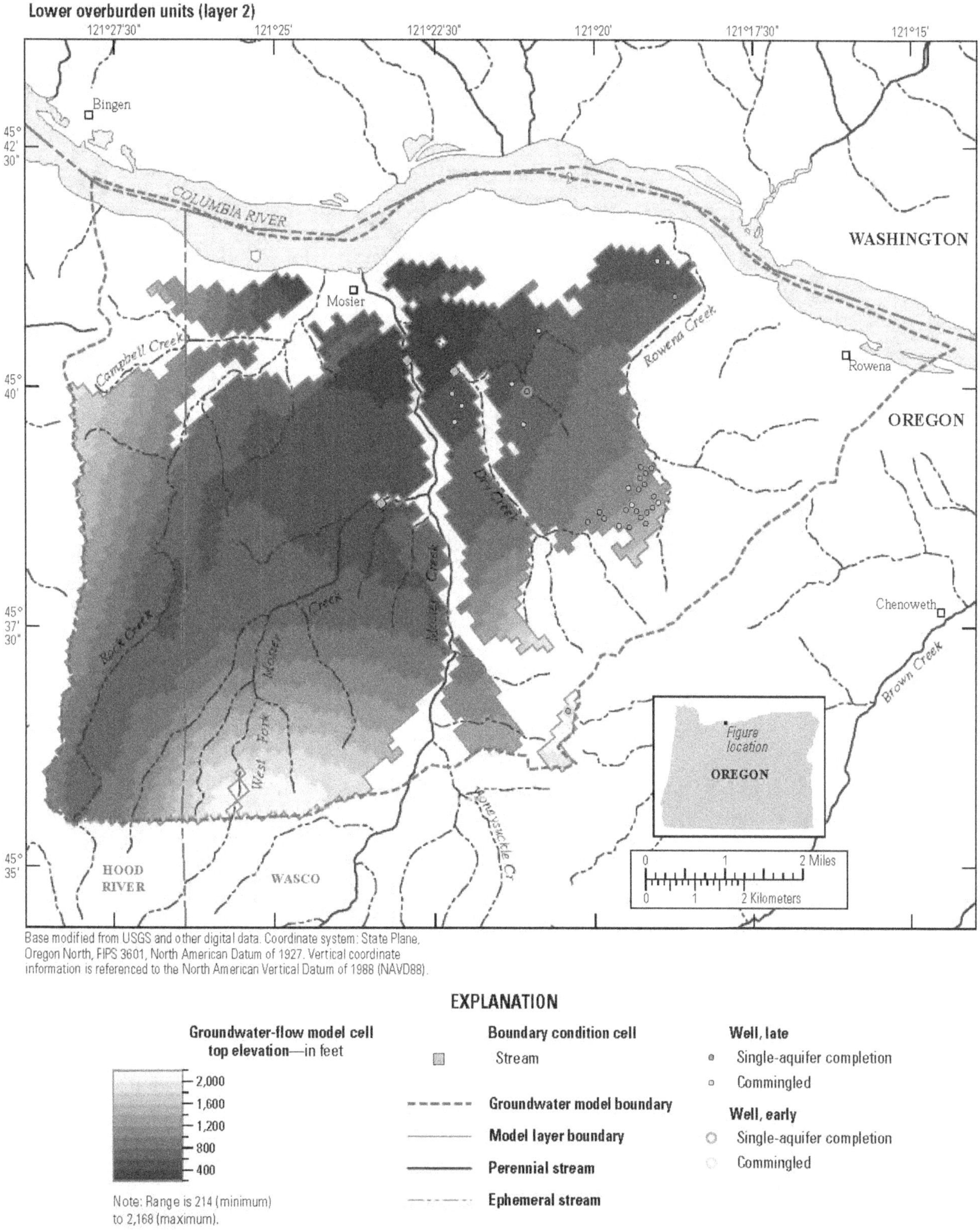

Lower overburden units (layer 2)
121°27'30"
121°25'
121°22'30"
121°20'
121°17'30"
121°15'
45° 42' 30"
45° 40'
45° 37' 30"
45° 35'
Bingen
COLUMBIA RIVER
WASHINGTON
Mosier
Campbell Creek
Rowena Creek
Rowena
OREGON
Dry Creek
Chenoweth
Rock Creek
Mosier Creek
West Fork Mosier Creek
Brown Creek
Honeysuckle Cr
Figure location
OREGON
HOOD RIVER
WASCO
0 1 2 Miles
0 1 2 Kilometers
Base modified from USGS and other digital data. Coordinate system: State Plane, Oregon North, FIPS 3601, North American Datum of 1927. Vertical coordinate information is referenced to the North American Vertical Datum of 1988 (NAVD88).
EXPLANATION
Groundwater-flow model cell top elevation—in feet
2,000
1,600
1,200
800
400
Note: Range is 214 (minimum) to 2,168 (maximum).
Boundary condition cell
Stream
Groundwater model boundary
Model layer boundary
Perennial stream
Ephemeral stream
Well, late
Single-aquifer completion
Commingled
Well, early
Single-aquifer completion
Commingled

Figure A10. The extent, model layer top elevation, and model boundary conditions of the lower overburden units (layer 2) in the Mosier, Oregon, groundwater-simulation model area.

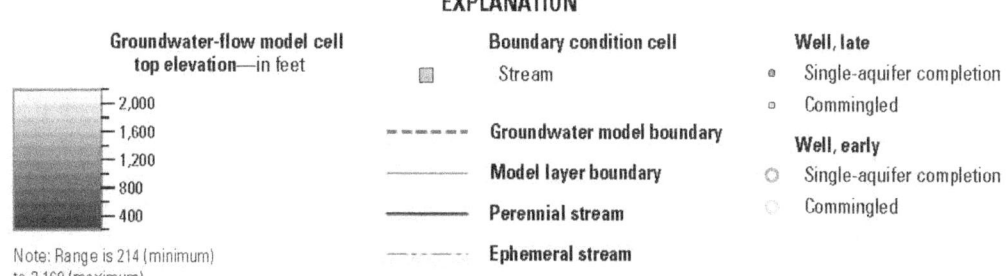

Pomona Basalt unit flow top [aquifer] (layer 3)

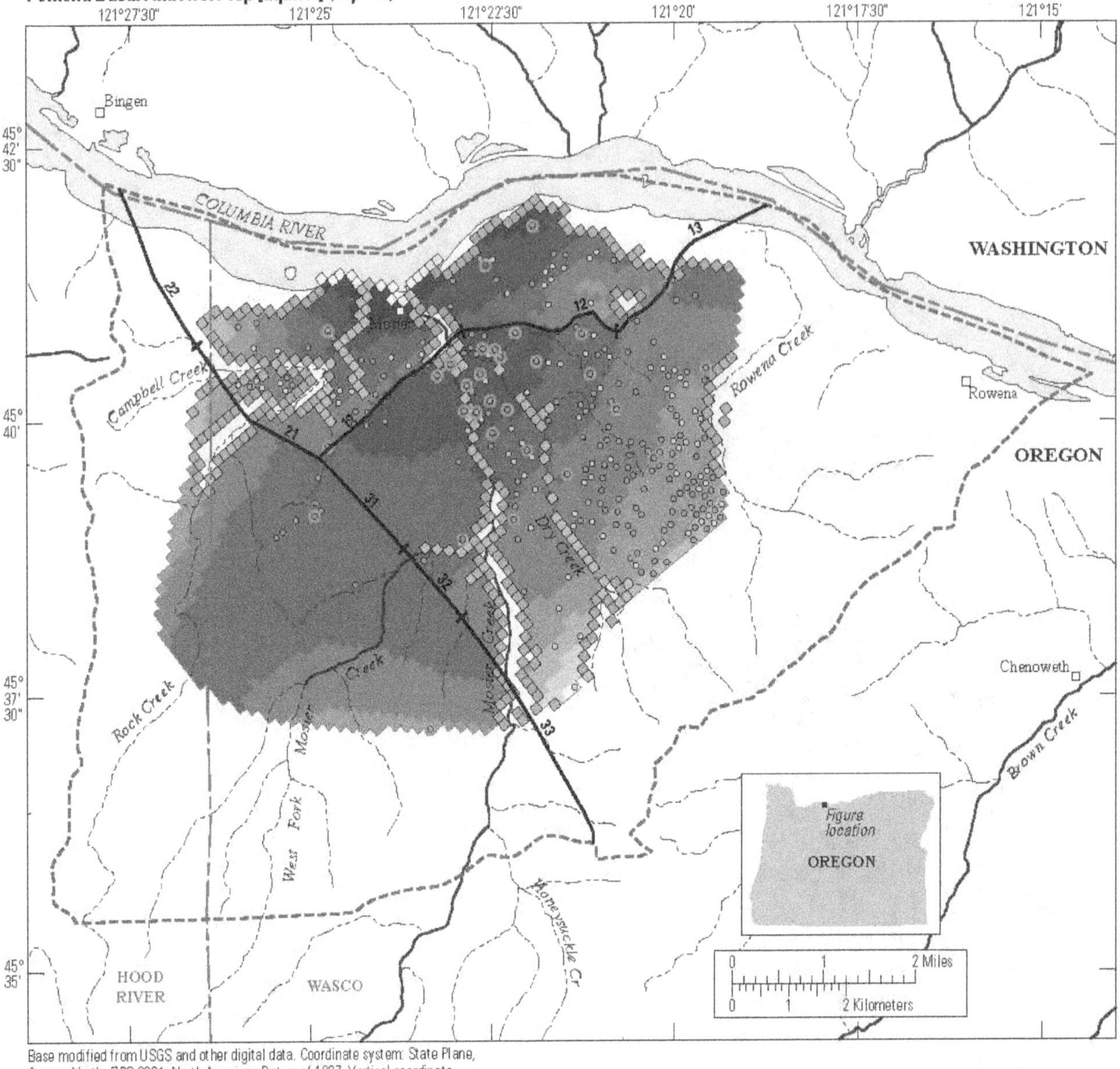

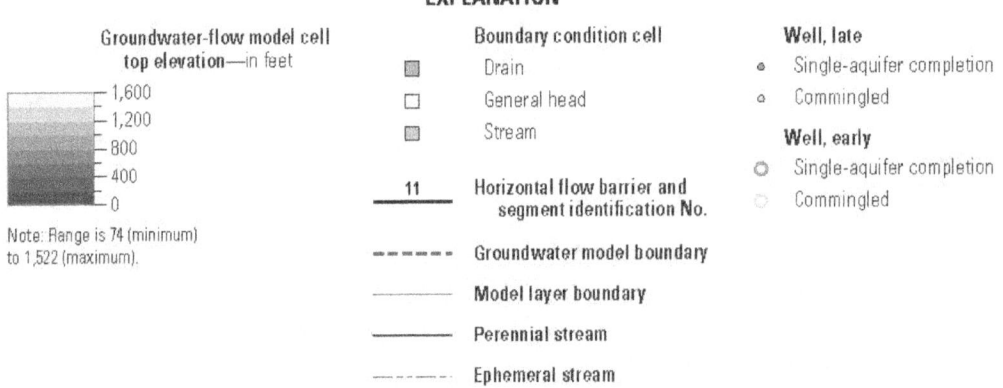

Base modified from USGS and other digital data. Coordinate system: State Plane,
Oregon North, FIPS 3601, North American Datum of 1927. Vertical coordinate
information is referenced to the North American Vertical Datum of 1988 (NAVD88).

EXPLANATION

Groundwater-flow model cell
top elevation—in feet

1,600
1,200
800
400
0

Note: Range is 74 (minimum)
to 1,522 (maximum).

Boundary condition cell

Drain

General head

Stream

11   Horizontal flow barrier and
segment identification No.

Groundwater model boundary

Model layer boundary

Perennial stream

Ephemeral stream

Well, late

Single-aquifer completion

Commingled

Well, early

Single-aquifer completion

Commingled

Figure A11.   The extent, model layer top elevation, and model boundary conditions of the Pomona Basalt unit flow top [aquifer]
(layer 3) in the Mosier, Oregon, groundwater-simulation model area.

Pomona Basalt unit flow interior [confining unit] (layer 4)

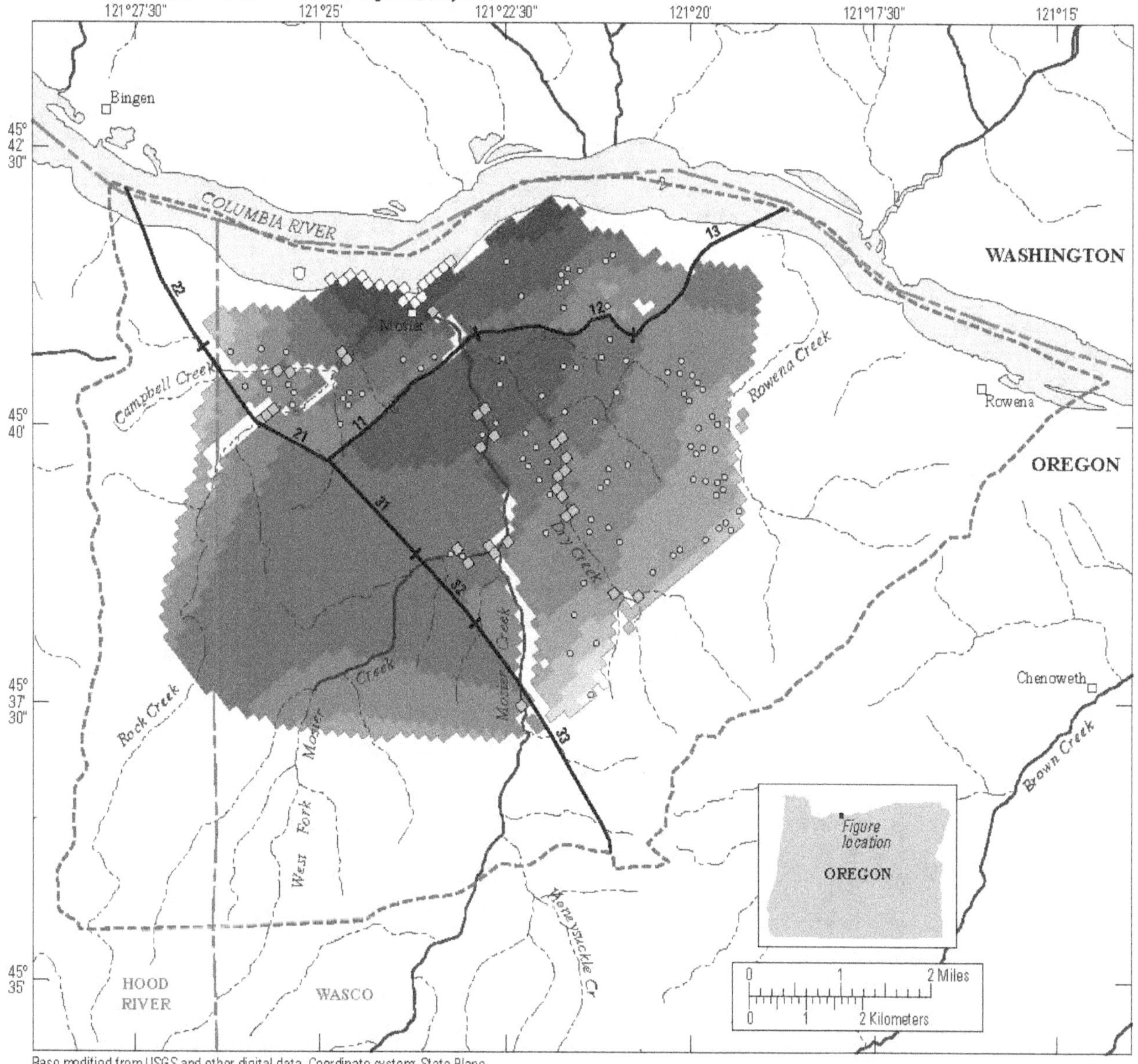

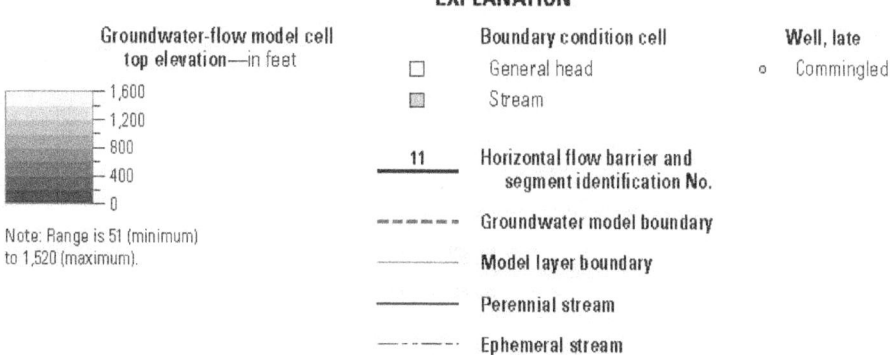

Base modified from USGS and other digital data. Coordinate system: State Plane, Oregon North, FIPS 3601, North American Datum of 1927. Vertical coordinate information is referenced to the North American Vertical Datum of 1988 (NAVD88).

EXPLANATION

Groundwater-flow model cell top elevation—in feet

1,600
1,200
800
400
0

Note: Range is 51 (minimum) to 1,520 (maximum).

Boundary condition cell
☐ General head
▨ Stream

—— 11 Horizontal flow barrier and segment identification No.

– – – – Groundwater model boundary

——— Model layer boundary

——— Perennial stream

–·–·– Ephemeral stream

Well, late
○ Commingled

Figure A12.   The extent, model layer top elevation, and model boundary conditions of the Pomona Basalt unit flow interior [confining unit] (layer 4) in the Mosier, Oregon, groundwater-simulation model area.

Pomona Basalt unit flow bottom [aquifer] (layer 5)

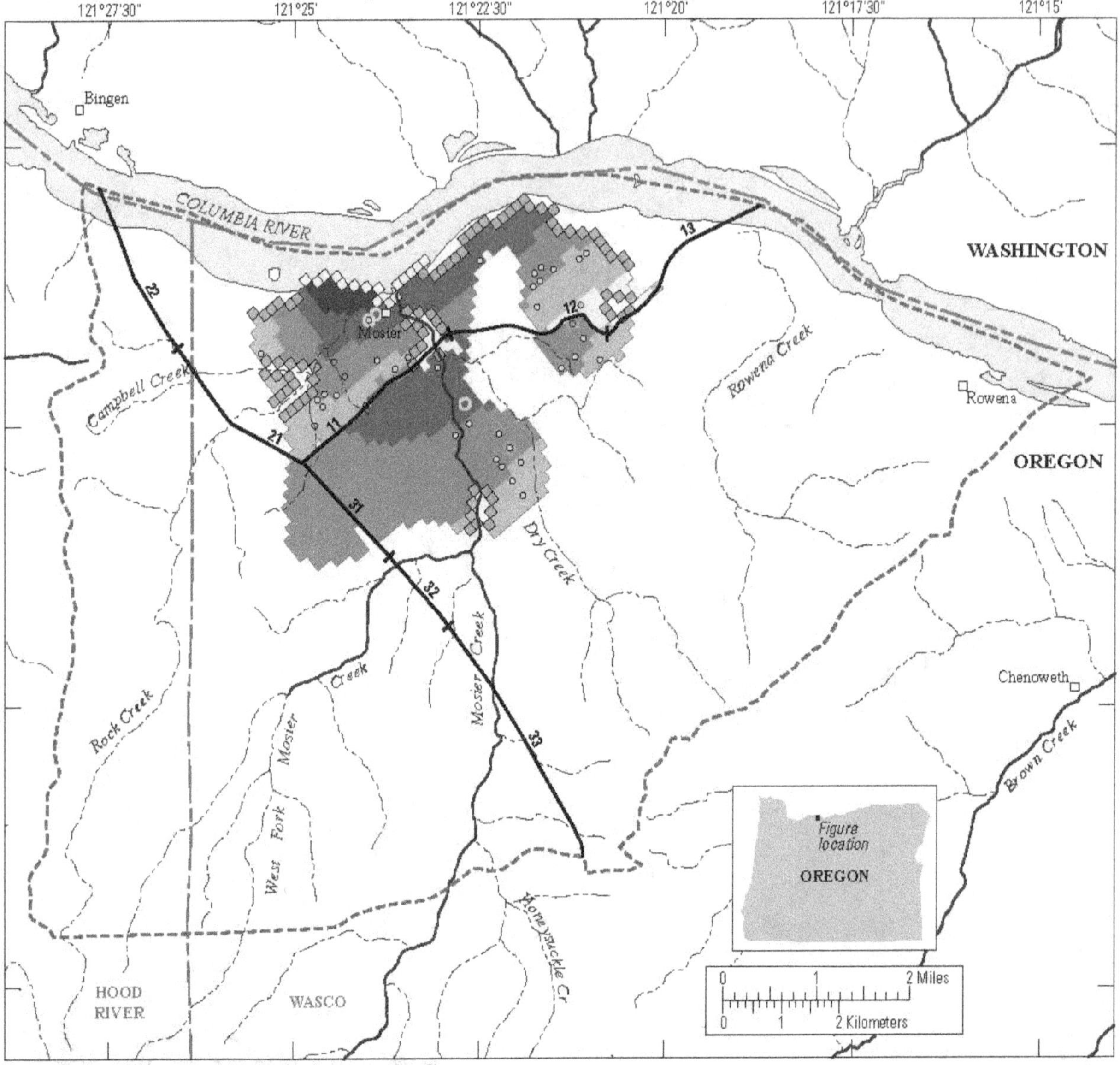

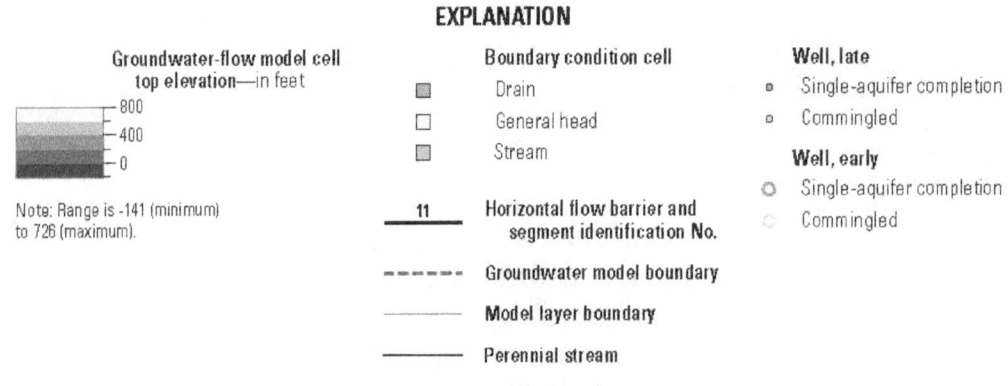

Base modified from USGS and other digital data. Coordinate system: State Plane, Oregon North, FIPS 3601, North American Datum of 1927. Vertical coordinate information is referenced to the North American Vertical Datum of 1988 (NAVD88).

EXPLANATION

Groundwater-flow model cell top elevation—in feet

800
400
0

Note: Range is -141 (minimum) to 726 (maximum).

Boundary condition cell

☐ Drain
☐ General head
☐ Stream

11 Horizontal flow barrier and segment identification No.

‒ ‒ ‒ ‒ Groundwater model boundary

———— Model layer boundary

———— Perennial stream

— · — · — Ephemeral stream

Well, late

◉ Single-aquifer completion
◎ Commingled

Well, early

○ Single-aquifer completion
○ Commingled

Figure A13.   The extent, model layer top elevation, and model boundary conditions of the Pomona Basalt unit flow bottom [aquifer] (layer 5) in the Mosier, Oregon, groundwater-simulation model area.

Selah interbed unit [confining unit] (layer 6)

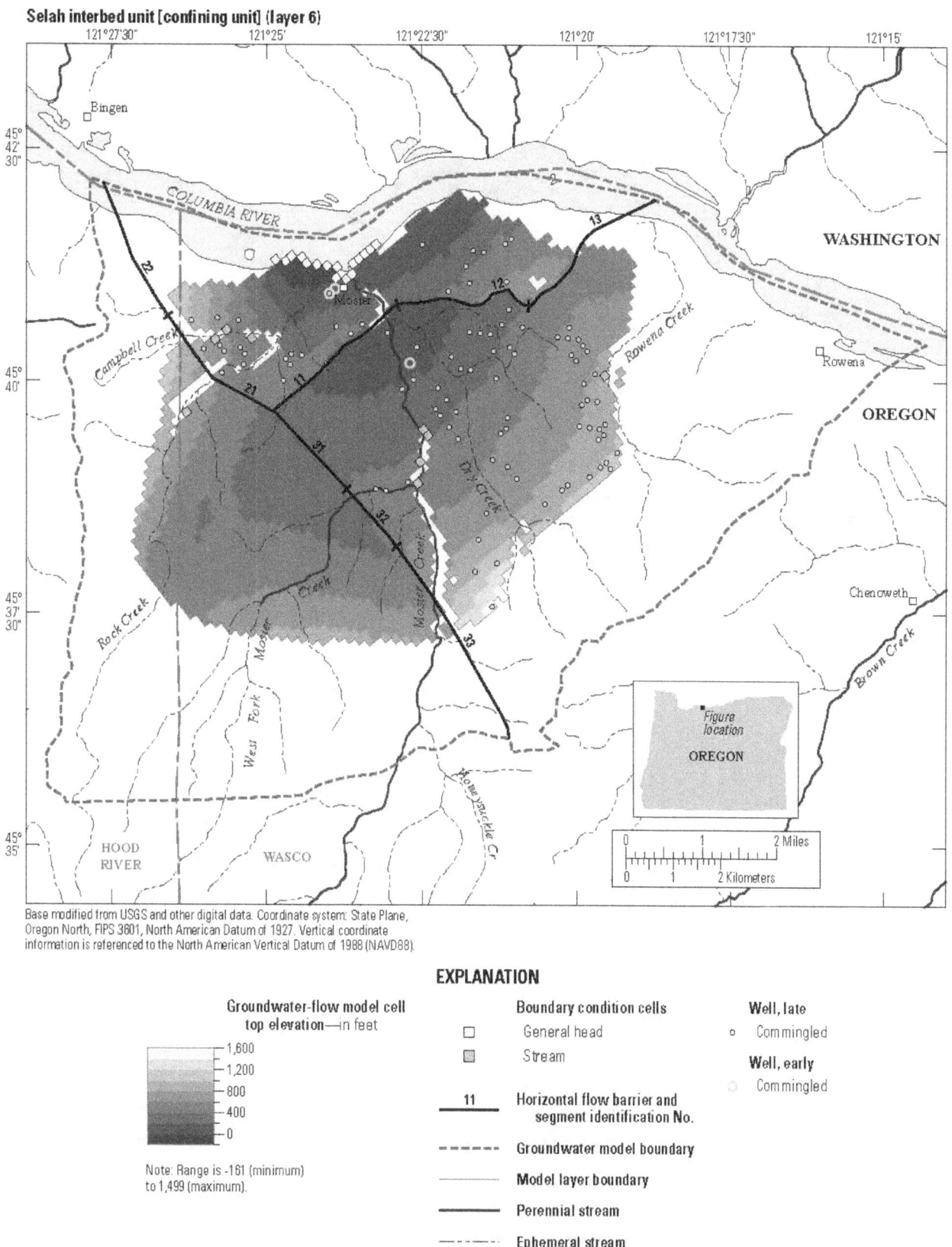

Base modified from USGS and other digital data. Coordinate system: State Plane,
Oregon North, FIPS 3601, North American Datum of 1927. Vertical coordinate
information is referenced to the North American Vertical Datum of 1988 (NAVD88).

EXPLANATION

**Groundwater-flow model cell
top elevation—in feet**

- 1,600
- 1,200
- 800
- 400
- 0

Note: Range is -161 (minimum)
to 1,499 (maximum).

Boundary condition cells

☐ General head

▨ Stream

11 Horizontal flow barrier and
segment identification No.

- - - - - Groundwater model boundary

———— Model layer boundary

——— Perennial stream

— · — · Ephemeral stream

Well, late

○ Commingled

Well, early

○ Commingled

Figure A14. The extent, model layer top elevation, and model boundary conditions of the Selah interbed unit [confining unit]
(layer 6) in the Mosier, Oregon, groundwater-simulation model area.

Lolo Basalt unit flow top [aquifer] (layer 7)

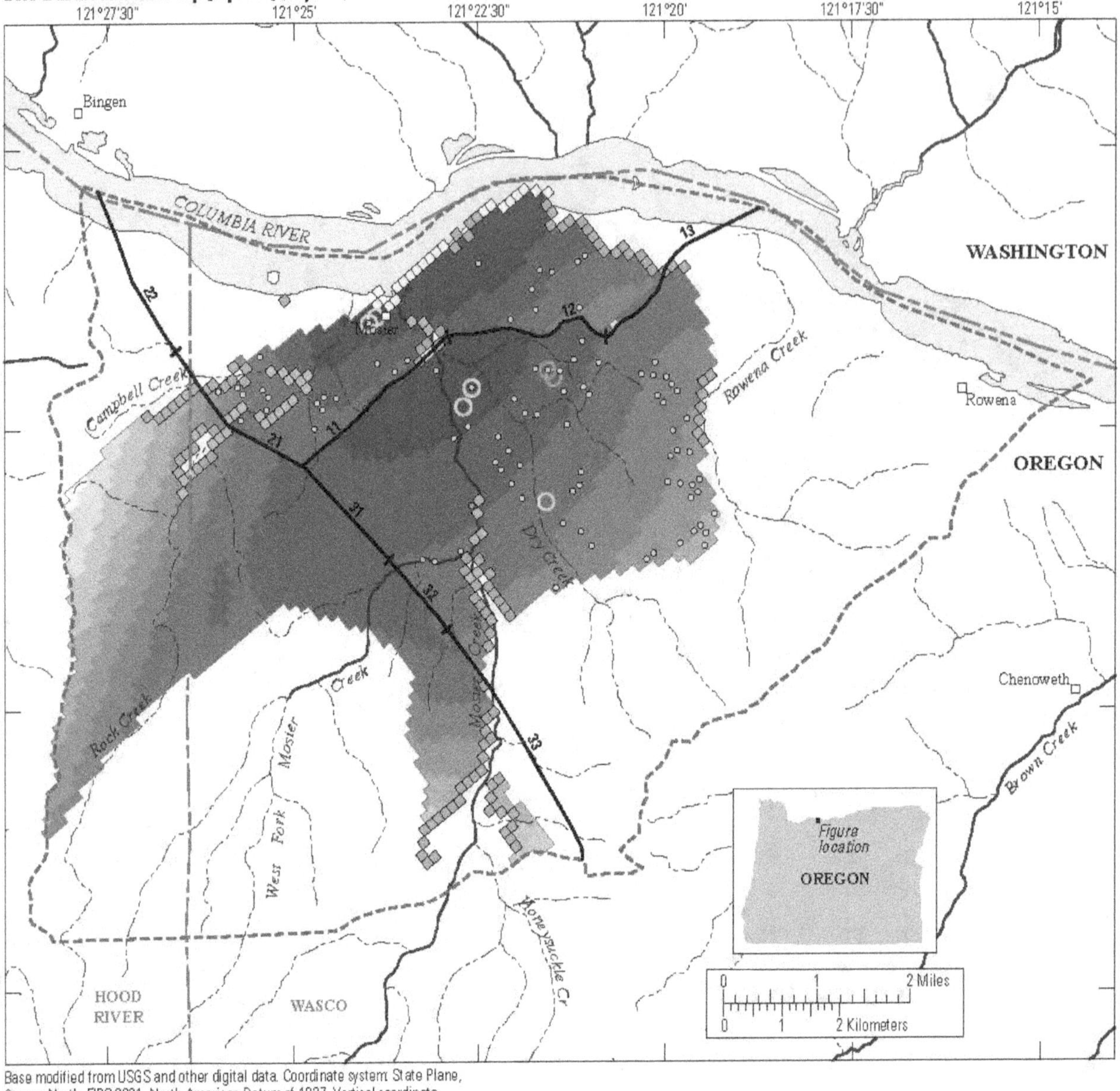

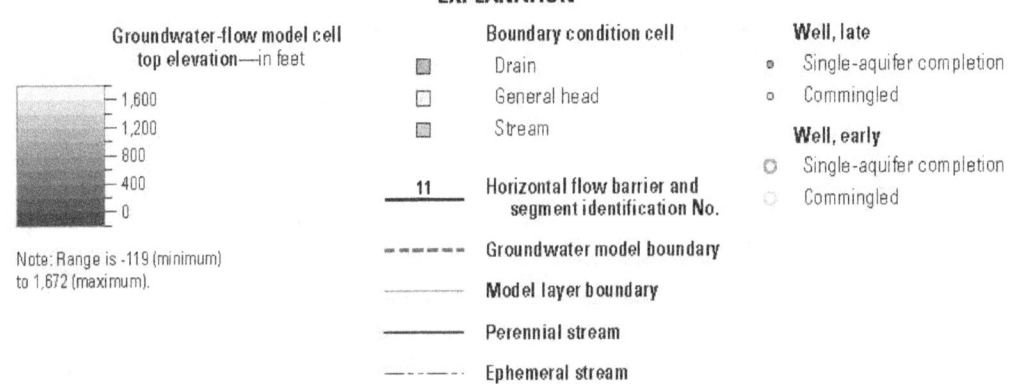

Base modified from USGS and other digital data. Coordinate system: State Plane,
Oregon North, FIPS 3601, North American Datum of 1927. Vertical coordinate
information is referenced to the North American Vertical Datum of 1988 (NAVD 88).

EXPLANATION

Groundwater-flow model cell
top elevation—in feet

1,600
1,200
800
400
0

Note: Range is -119 (minimum)
to 1,672 (maximum).

Boundary condition cell

Drain

General head

Stream

11   Horizontal flow barrier and
segment identification No.

Groundwater model boundary

Model layer boundary

Perennial stream

Ephemeral stream

Well, late

o   Single-aquifer completion

o   Commingled

Well, early

⊙   Single-aquifer completion

◌   Commingled

Figure A15. The extent, model layer top elevation, and model boundary conditions of the Lolo Basalt unit flow top [aquifer] (layer 7)
in the Mosier, Oregon, groundwater-simulation model area.

Lolo Basalt unit flow interior [confining unit] (layer 8)

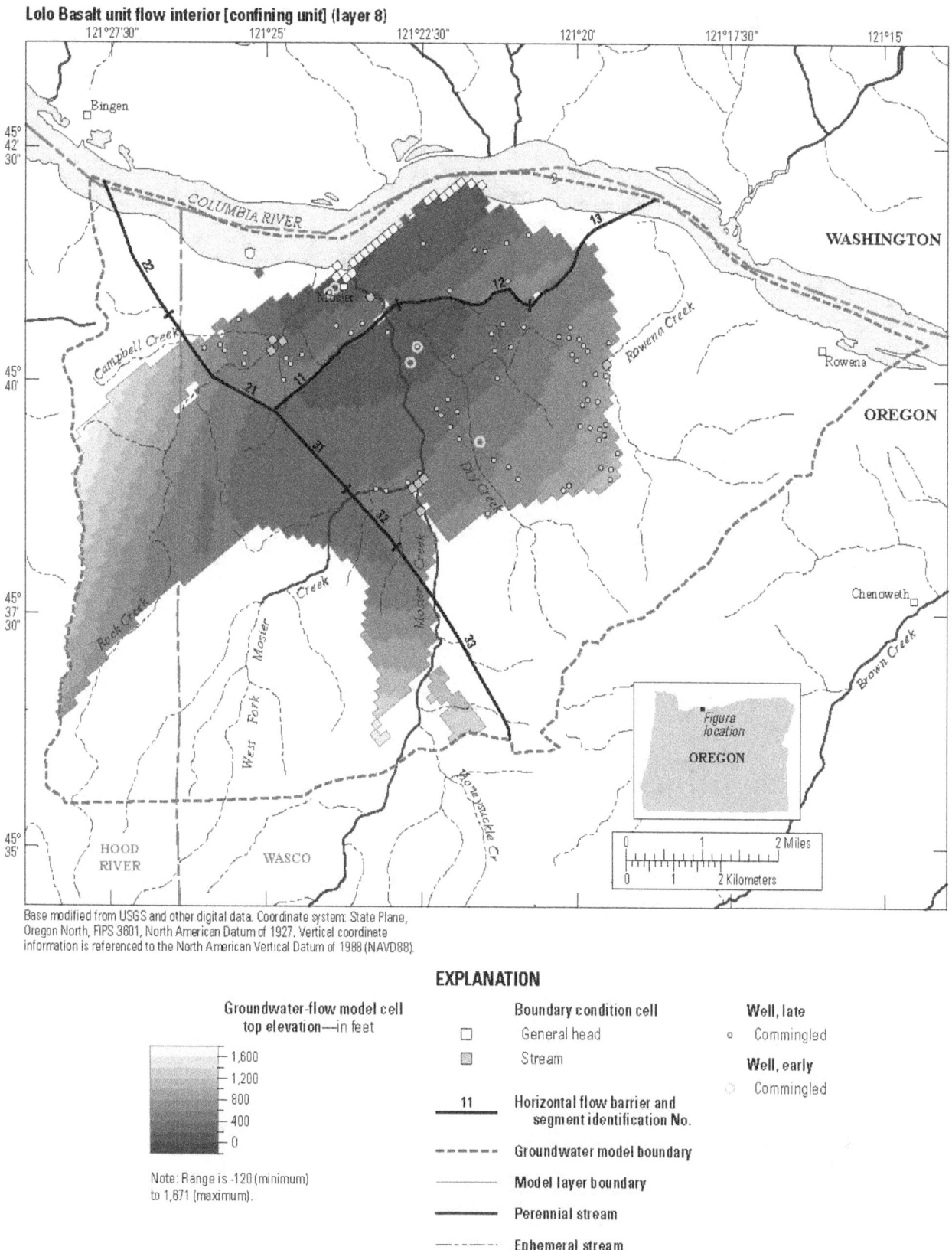

Base modified from USGS and other digital data. Coordinate system: State Plane, Oregon North, FIPS 3601, North American Datum of 1927. Vertical coordinate information is referenced to the North American Vertical Datum of 1988 (NAVD88).

EXPLANATION

Groundwater-flow model cell top elevation—in feet

1,600
1,200
800
400
0

Note: Range is -120 (minimum) to 1,671 (maximum).

Boundary condition cell

☐ General head

▨ Stream

—11— Horizontal flow barrier and segment identification No.

- - - - Groundwater model boundary

——— Model layer boundary

——— Perennial stream

—·—·— Ephemeral stream

Well, late
○ Commingled

Well, early
○ Commingled

Figure A16.   The extent, model layer top elevation, and model boundary conditions of the Lolo Basalt unit flow interior [confining unit] (layer 8) in the Mosier, Oregon, groundwater-simulation model area.

Rosalia Basalt unit flow top [aquifer] (layer 9)

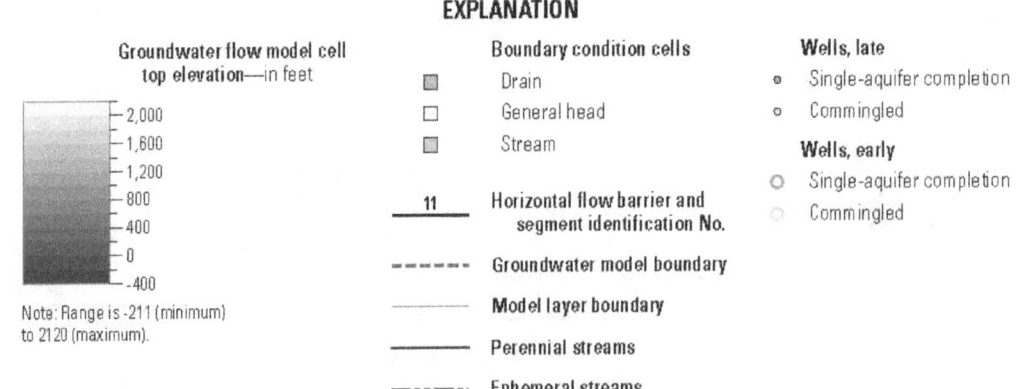

121°27'30"
121°25'
121°22'30"
121°20'
121°17'30"
121°15'
45° 42' 30"
45° 40'
45° 37' 30"
45° 35'
Bingen
COLUMBIA RIVER
WASHINGTON
OREGON
Mosier
Rowena
Rowena Creek
Rock Creek
Dry Creek
Mosier Creek
West Fork Mosier Creek
Brown Creek
Chenoweth
Honeysuckle Cr
HOOD RIVER
WASCO
22
21
11
12
13
31
32
33
Figure location
OREGON
0 1 2 Miles
0 1 2 Kilometers
Base modified from USGS and other digital data. Coordinate system: State Plane, Oregon North, FIPS 3601, North American Datum of 1927. Vertical coordinate information is referenced to the North American Vertical Datum of 1988 (NAVD 88).
EXPLANATION
Groundwater flow model cell top elevation—in feet
2,000
1,600
1,200
800
400
0
-400
Note: Range is -211 (minimum) to 2120 (maximum).
Boundary condition cells
Drain
General head
Stream
11
Horizontal flow barrier and segment identification No.
Groundwater model boundary
Model layer boundary
Perennial streams
Ephemeral streams
Wells, late
Single-aquifer completion
Commingled
Wells, early
Single-aquifer completion
Commingled

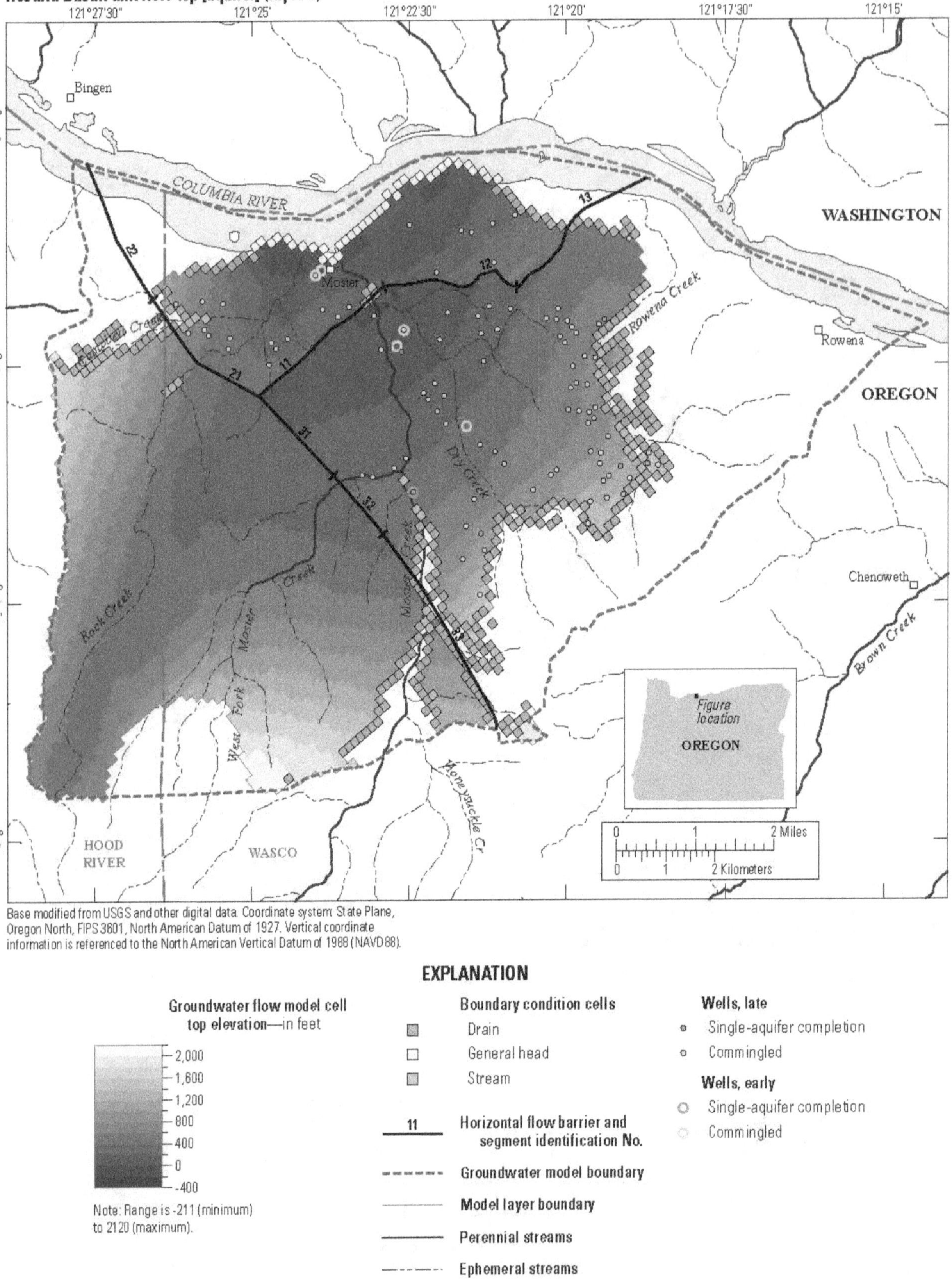

Figure A17. The extent, model layer top elevation, and model boundary conditions of the Rosalia Basalt unit flow top [aquifer] (layer 9) in the Mosier, Oregon, groundwater-simulation model area.

Rosalia Basalt unit flow interior [confining unit] (layer 10)

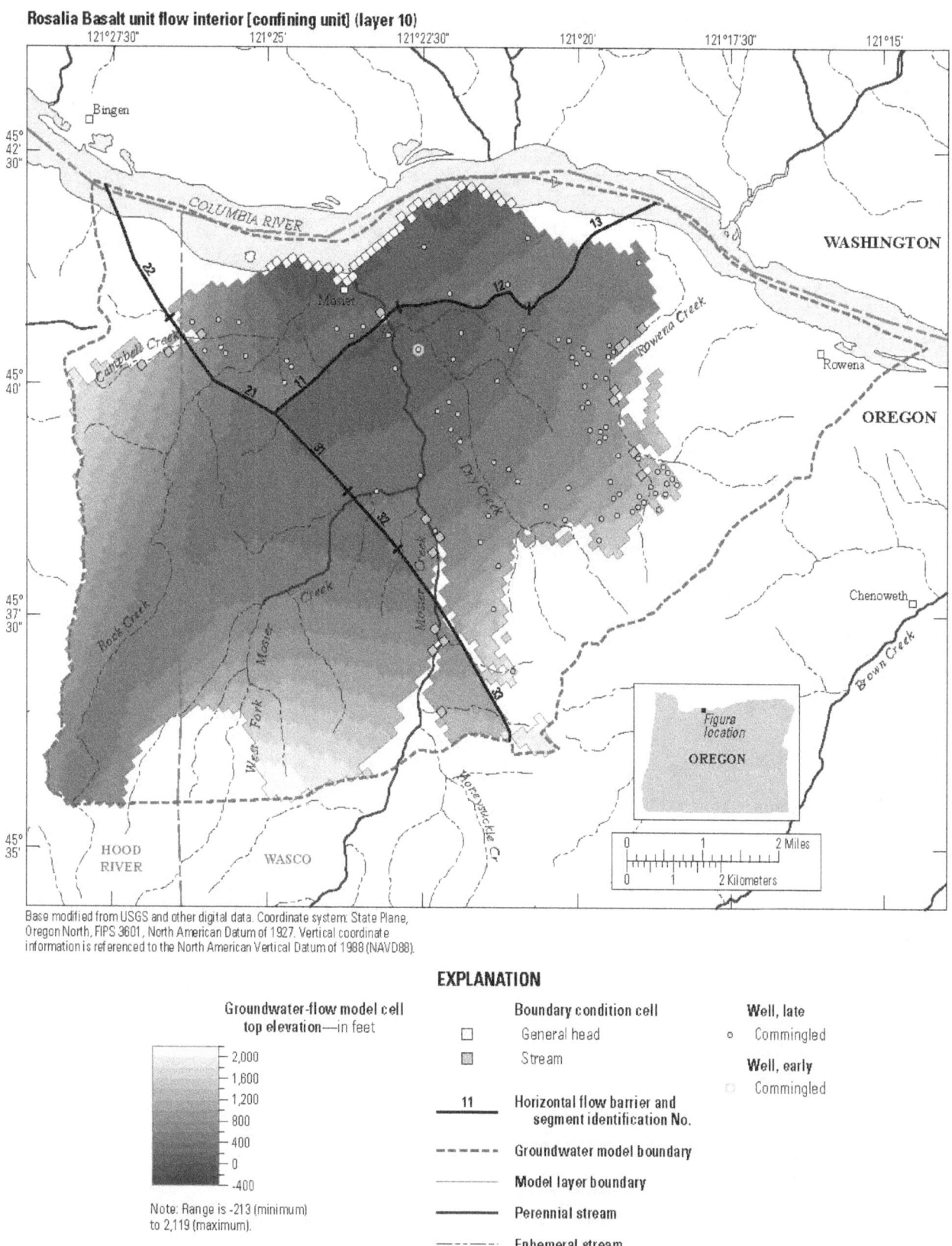

Base modified from USGS and other digital data. Coordinate system: State Plane,
Oregon North, FIPS 3601, North American Datum of 1927. Vertical coordinate
information is referenced to the North American Vertical Datum of 1988 (NAVD88).

EXPLANATION

Groundwater-flow model cell
top elevation—in feet

2,000
1,600
1,200
800
400
0
-400

Note: Range is -213 (minimum)
to 2,119 (maximum).

Boundary condition cell

☐   General head

▨   Stream

—11—   Horizontal flow barrier and
         segment identification No.

- - -   Groundwater model boundary

——   Model layer boundary

——   Perennial stream

—·—·—   Ephemeral stream

Well, late
○   Commingled

Well, early
○   Commingled

Figure A18.   The extent, model layer top elevation, and model boundary conditions of the Rosalia Basalt unit flow interior
[confining unit] (layer 10) in the Mosier, Oregon, groundwater-simulation model area.

Quincy-Squaw Creek interbed unit [confining unit] (layer 11)

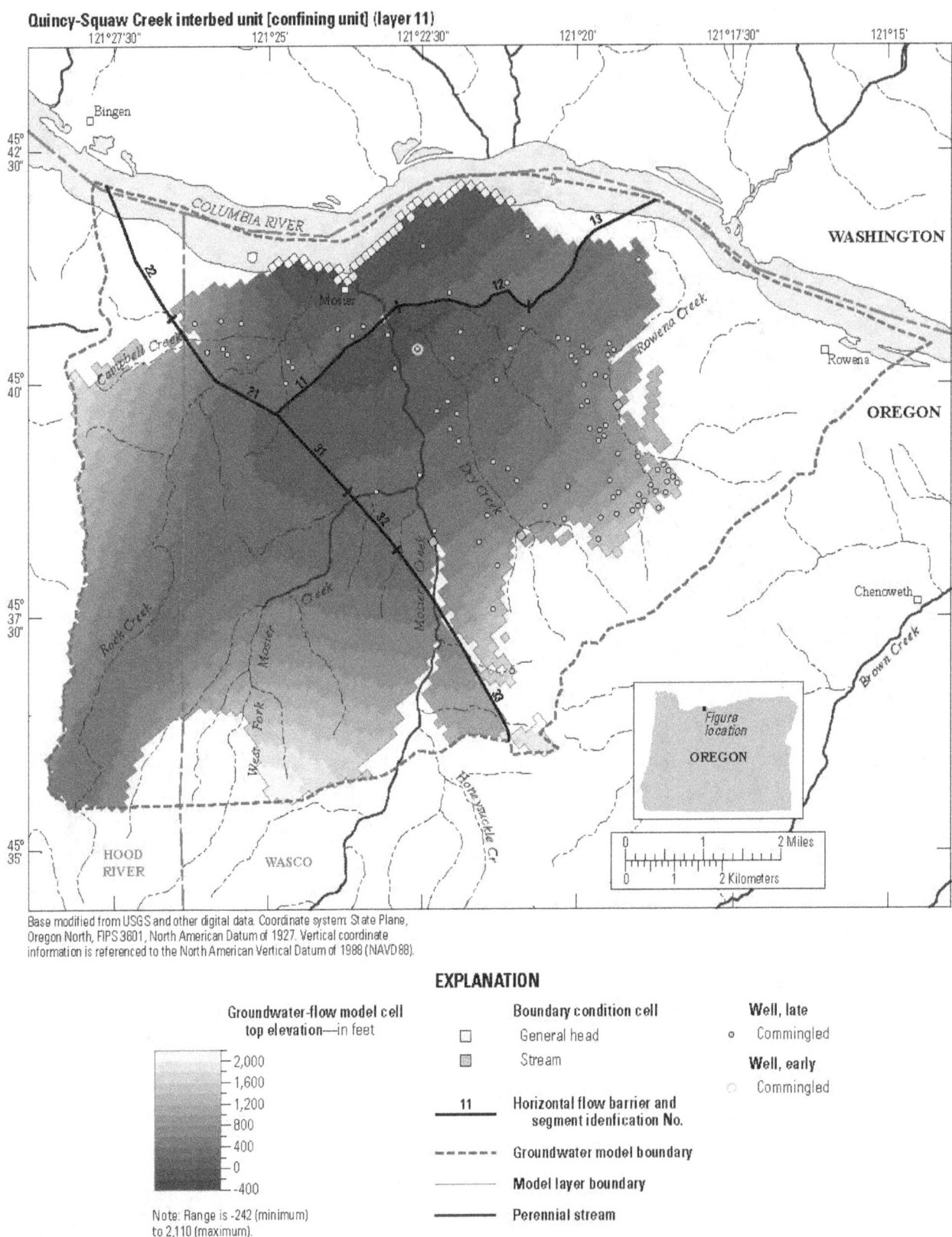

Base modified from USGS and other digital data. Coordinate system: State Plane,
Oregon North, FIPS 3601, North American Datum of 1927. Vertical coordinate
information is referenced to the North American Vertical Datum of 1988 (NAVD 88).

EXPLANATION

Groundwater-flow model cell
top elevation—in feet

- 2,000
- 1,600
- 1,200
- 800
- 400
- 0
- -400

Note: Range is -242 (minimum)
to 2,110 (maximum).

Boundary condition cell
- □ General head
- ▨ Stream

—11— Horizontal flow barrier and
segment idenfication No.

- – – – Groundwater model boundary
- —— Model layer boundary
- —— Perennial stream
- – · – · Ephemeral stream

Well, late
- ○ Commingled

Well, early
- ○ Commingled

Figure A19. The extent, model layer top elevation, and model boundary conditions of the Quincy-Squaw Creek interbed unit
[confining unit] (layer 11) in the Mosier, Oregon, groundwater-simulation model area.

Frenchman Springs Basalt unit flow top [aquifer] (layer 12)

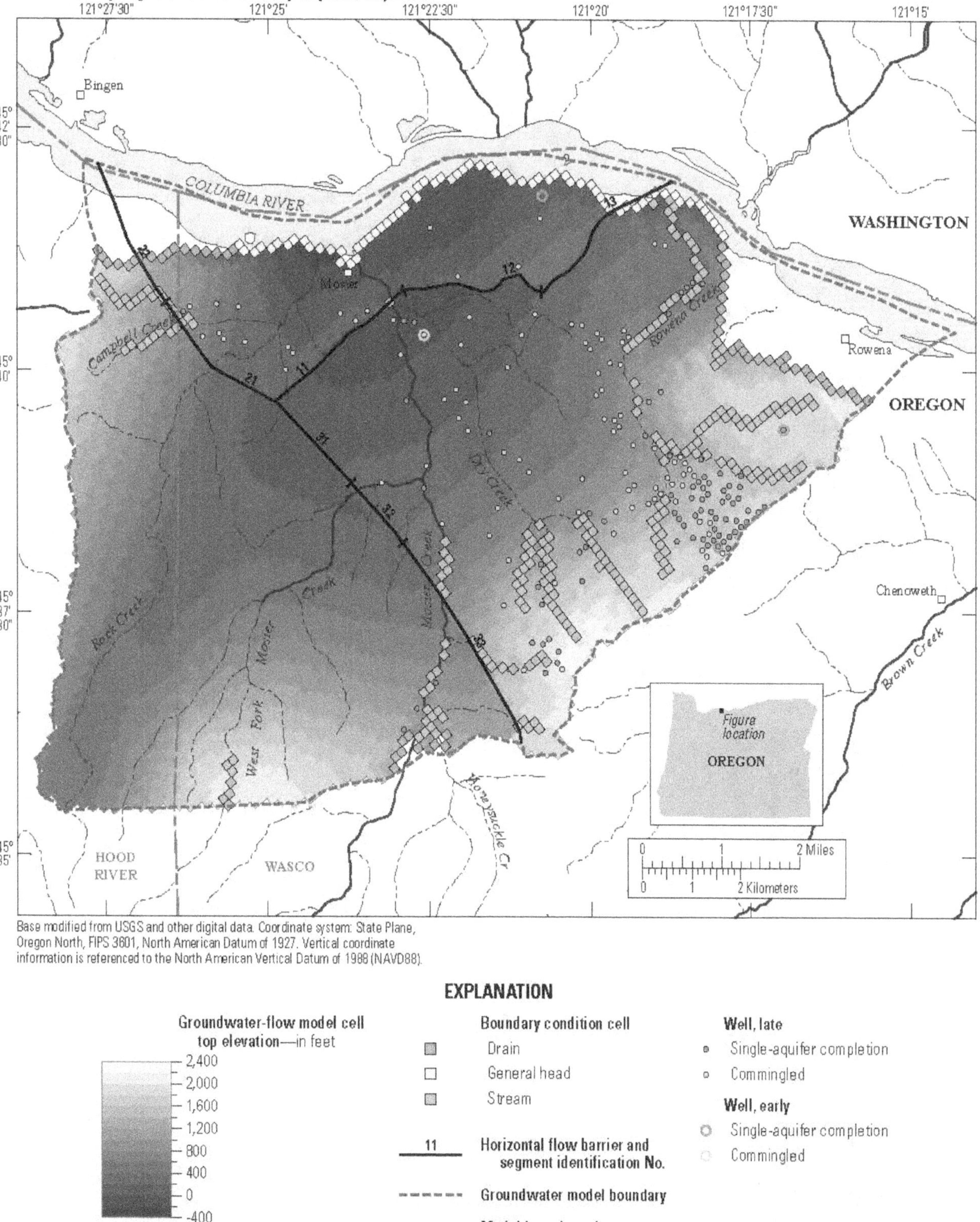

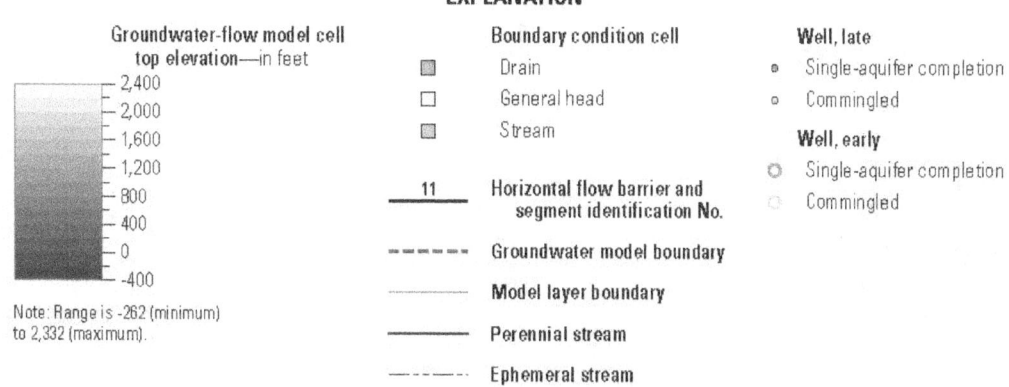

Base modified from USGS and other digital data. Coordinate system: State Plane,
Oregon North, FIPS 3601, North American Datum of 1927. Vertical coordinate
information is referenced to the North American Vertical Datum of 1988 (NAVD88).

EXPLANATION

**Groundwater-flow model cell
top elevation—in feet**

2,400
2,000
1,600
1,200
800
400
0
-400

Note: Range is -262 (minimum)
to 2,332 (maximum).

Boundary condition cell

▨ Drain

☐ General head

▨ Stream

11 Horizontal flow barrier and
    segment identification No.

- - - - Groundwater model boundary

——— Model layer boundary

——— Perennial stream

—·—·— Ephemeral stream

Well, late

⊙ Single-aquifer completion

○ Commingled

Well, early

⊚ Single-aquifer completion

○ Commingled

Figure A20.   The extent, model layer top elevation, and model boundary conditions of the Frenchman Springs Basalt unit flow top
[aquifer] (layer 12) in the Mosier, Oregon, groundwater-simulation model area.

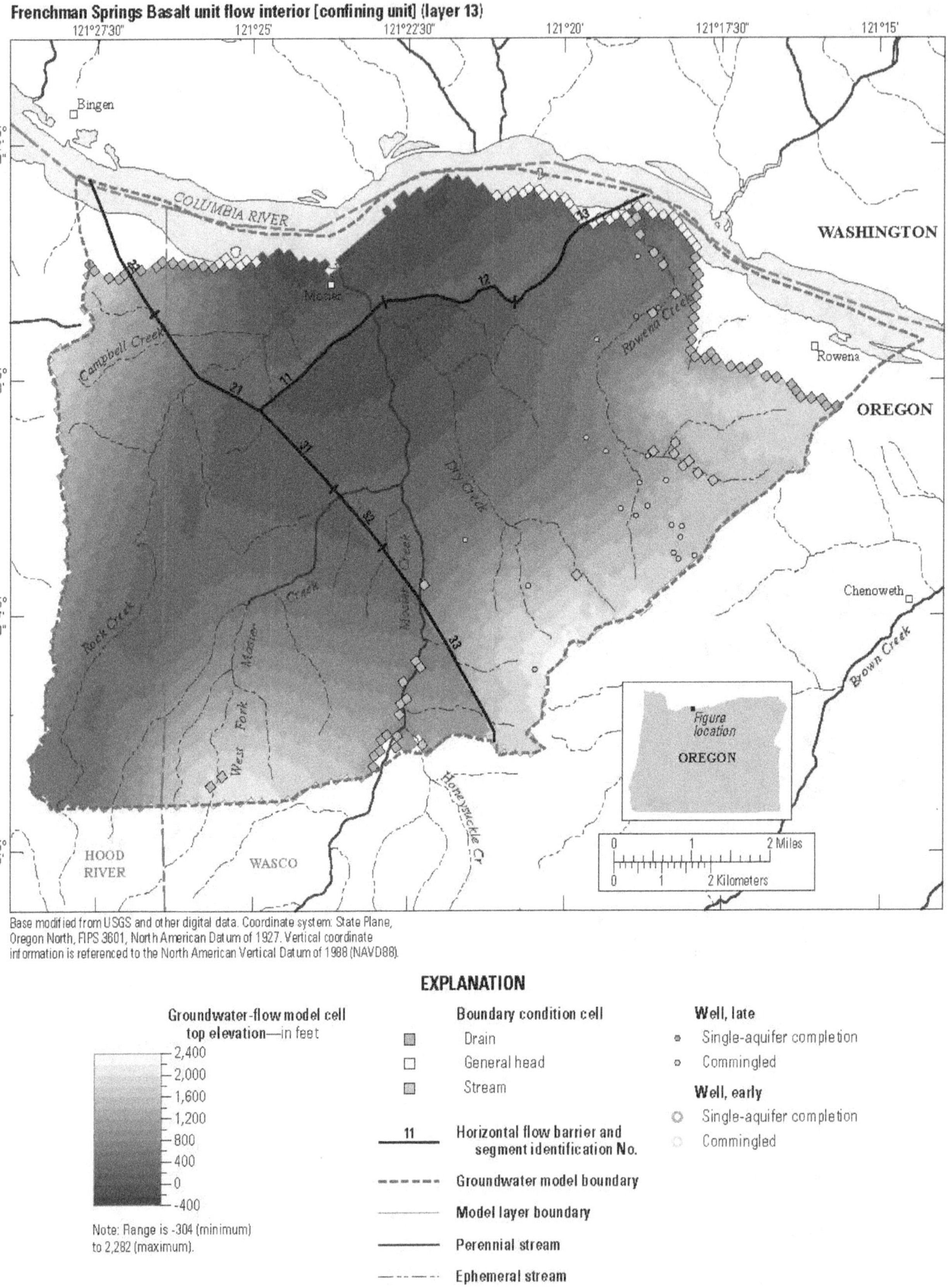

Frenchman Springs Basalt unit flow interior [confining unit] (layer 13)

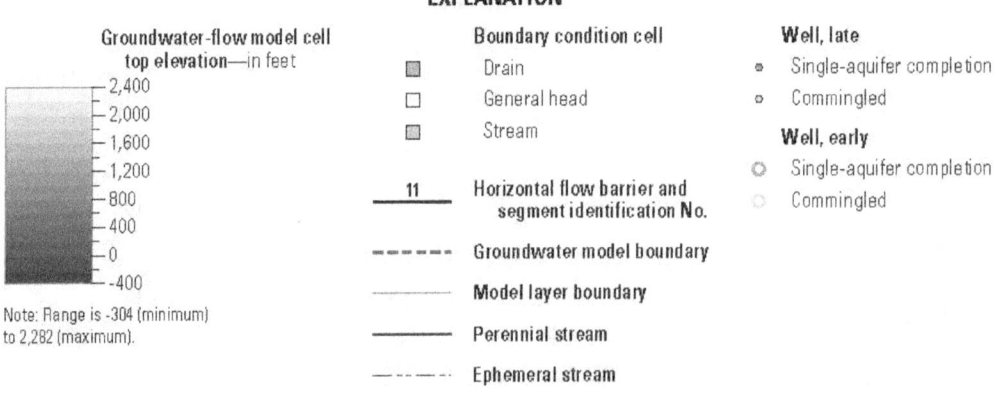

Base modified from USGS and other digital data. Coordinate system: State Plane,
Oregon North, FIPS 3601, North American Datum of 1927. Vertical coordinate
information is referenced to the North American Vertical Datum of 1988 (NAVD88).

EXPLANATION

Groundwater-flow model cell
top elevation—in feet

- 2,400
- 2,000
- 1,600
- 1,200
- 800
- 400
- 0
- -400

Note: Range is -304 (minimum)
to 2,282 (maximum).

Boundary condition cell

▨ Drain
☐ General head
▨ Stream

——11—— Horizontal flow barrier and
segment identification No.

– – – Groundwater model boundary

——— Model layer boundary

——— Perennial stream

—·—·— Ephemeral stream

Well, late

◉ Single-aquifer completion
◎ Commingled

Well, early

◉ Single-aquifer completion
◎ Commingled

Figure A21.   The extent, model layer top elevation, and model boundary conditions of the Frenchman Springs Basalt unit flow interior
[confining unit] (layer 13) in the Mosier, Oregon, groundwater-simulation model area.

Grande Ronde Basalt unit flow top [aquifer] (layer 14)

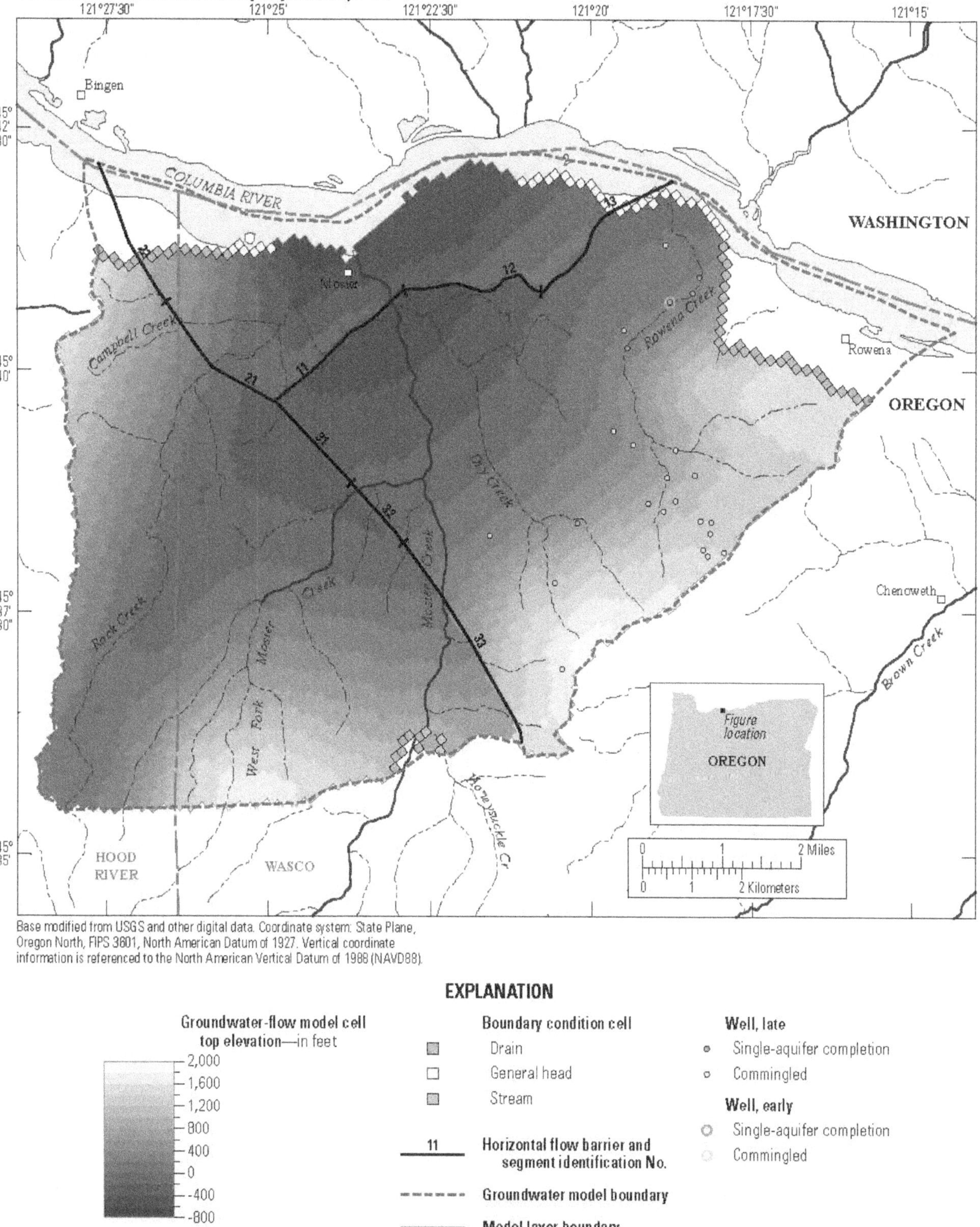

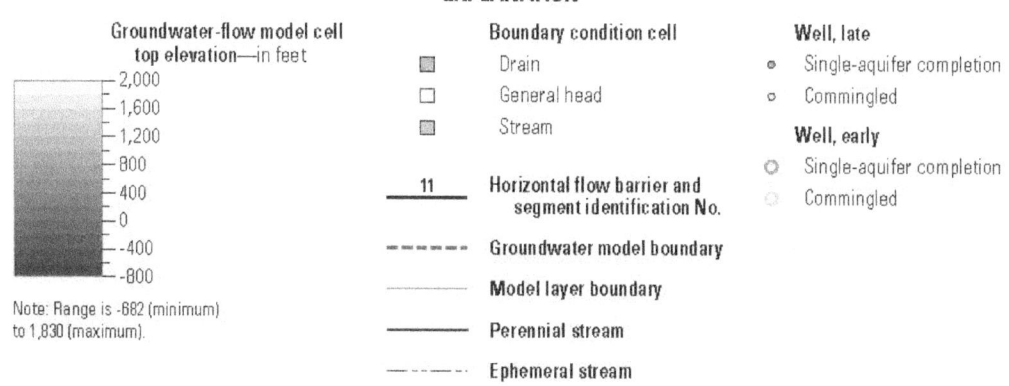

Base modified from USGS and other digital data. Coordinate system: State Plane, Oregon North, FIPS 3601, North American Datum of 1927. Vertical coordinate information is referenced to the North American Vertical Datum of 1988 (NAVD88).

EXPLANATION

Groundwater-flow model cell top elevation—in feet

2,000
1,600
1,200
800
400
0
-400
-800

Note: Range is -682 (minimum) to 1,830 (maximum).

Boundary condition cell

▨ Drain
☐ General head
▨ Stream

—— 11 —— Horizontal flow barrier and segment identification No.

– – – – Groundwater model boundary

———— Model layer boundary

———— Perennial stream

– · – · – Ephemeral stream

Well, late

⊙ Single-aquifer completion
○ Commingled

Well, early

◎ Single-aquifer completion
○ Commingled

Figure A22. The extent, model layer top elevation, and model boundary conditions of the Grande Ronde Basalt unit flow top [aquifer] (layer 14) in the Mosier, Oregon, groundwater-simulation model area.

A.5—Elevation of Tops of Geologic Model Units in Wells in the Mosier, Oregon, Study Area

Table A3 contains all well log interpretations used to construct the geologic model and the source of the interpretation.

Table A3.   Elevation of the tops of hydrogeologic units in wells in the Mosier, Oregon, study area.

[This table is available as a digital spreadsheet at http://pubs.usgs.gov/sir/2012/5002. **State plane:** State plane Oregon North (North American Datum of 1927) northing and easting coordinates (Kienle, 1995; Jervey, 1996). **Land-surface elevation:** Determined using state plane northing and easting coordinates (Kienle, 1995; Jervey, 1996) and USGS 10 meter digital elevation model (DEM), North American Vertical Datum of 1988 (NAVD 88). **Elevations:** Determined using well log stratigraphic interpretations (Kienle, 1995; Jervey, 1996) and land-surface elevation, NAVD 88. Grey italicized elevations have been interpreted below the drilled well hole depths using cross-sectional maps (Jervey, 1996). **Abbreviations:** USGS, U.S. Geological Survey; –, no dat]

Township range and section	County well No.	State well No.	USGS well No.	Station name	State plane Easting (feet)	Northing (feet)	Land-surface elevation (feet)	Alluvium deposits (feet)	Glaciofluvial deposits (feet)	Chenoweth Formation (feet)	Pomona Member (feet)	Selah Interbed (feet)	Lolo Priest Rapids Member (feet)	Rosalia Priest Rapids Member (feet)	Quincy-Squaw Creek Interbed (feet)	Frenchman Springs Member (feet)	Grande Ronde Formation (feet)	Bottom elevation (feet)	Stratigraphic interpretation source
2N11E01	467	WASC 2730	—	—	1775250	738750	320	—	320	—	296	110	101	58	—	-85	—	-175	Keinle
2N11E02	466	WASC 2732	45405312124180 1	02N/11E-02DCA1	1768651	735637	152	152	—	—	—	130	120	—	—	-70	—	-120	Jervey
2N11E02	468	WASC 2733	45405712124120 1	02N/11E-02DDB1	1769191	736286	160	160	—	—	—	90	70	—	—	-100	—	-150	Jervey
2N11E02	469	WASC 2731	—	—	1767340	735556	359	359	—	—	314	—	154	—	—	-46	—	-71	Jervey
2N11E02	470	WASC 2734	45405812123590 1	02N/11E-02DAD1	1770047	736231	155	155	—	—	—	87	77	—	-93	-120	—	-170	Jervey
2N11E02	471	—	—	—	1765529	735781	576	—	—	—	576	—	419	—	—	231	—	181	Jervey
2N11E03	473	—	—	—	1762720	736080	917	—	—	—	917	—	782	—	592	562	—	422	Jervey
2N11E03	475	—	—	—	1760914	730425	686	—	—	—	—	—	686	—	—	386	—	336	Jervey
2N11E10	478	WASC 2744	45401912125410 1	02N/11E-10ACC1	1762973	732357	820	—	—	—	820	—	660	—	—	420	—	390	Jervey
2N11E10	479	WASC 2745	45402012125260 1	02N/11E-10ADC1	1763810	732054	770	—	—	—	770	640	620	—	470	—	—	450	Jervey
2N11E10	483	WASC 2738	—	—	1762809	731155	683	—	—	—	—	—	683	—	—	358	—	308	Jervey
2N11E10	513	WASC 2777	—	—	1764651	732305	660	—	—	—	660	—	530	—	—	350	—	300	Jervey
2N11E10	1204	WASC 2741	45404212125270 1	02N/11E-10AAB1	1763507	734433	750	—	—	—	750	—	605	—	—	435	—	325	Jervey
2N11E10	1898	WASC 1898	—	—	1764295	732331	727	—	—	—	727	—	572	—	—	370	—	177	Jervey
2N11E10	2036	WASC 2036	—	—	1762256	734591	996	—	—	996	906	—	756	—	—	526	—	446	Jervey
2N11E10	2070	WASC 2070	—	—	1762539	732717	936	—	—	936	876	—	764	—	—	464	—	424	Jervey
2N11E10	2138	WASC 2138	—	—	1761395	732749	1,058	—	—	—	1,058	—	973	—	—	818	—	613	Jervey
2N11E11	487	WASC 2752	—	—	1768940	731420	690	—	—	—	690	400	305	195	—	—	—	165	Keinle
2N11E11	488	WASC 2751	—	—	1768496	732050	621	—	—	—	621	470	360	—	—	230	—	180	Jervey
2N11E11	490	WASC 2750	—	—	1767500	732250	555	—	—	555	506	325	320	—	—	160	—	150	Keinle
2N11E11	491	WASC 2753	45401612124400 1	02N/11E-11CAA1	1766890	731962	531	—	—	531	500	331	323	—	—	171	—	121	Keinle
2N11E11	492	WASC 2754	45400112124400 1	02.00N/11.00E-11CDA01	1767100	730520	737	—	—	—	737	567	555	—	—	387	—	309	Keinle
2N11E11	493	WASC 2771	—	—	1769665	731401	749	—	—	749	349	—	169	—	—	—	—	119	Keinle
2N11E11	1871	—	—	—	1768334	731098	688	—	—	688	640	475	450	—	325	310	—	40	Jervey
2N11E11	1872	—	—	—	1767649	730385	780	—	—	780	590	—	410	—	—	240	—	190	Jervey
2N11E11	2022	WASC 2022	—	—	1767502	731512	605	—	—	605	570	—	430	—	220	200	—	175	Jervey
2N11E12	2079	WASC 2079	—	—	1768200	732150	608	—	—	608	538	400	360	—	—	170	—	120	Keinle
2N11E12	494	WASC 2765	45403312123010 1	02N/11E-12ABD1	1774213	732372	299	—	299	—	289	69	29	—	—	-91	—	-101	Keinle
2N11E12	496	WASC 2759	45403112122400 1	02N/11E-12AAD1	1775532	733511	372	—	372	369	211	15	-100	—	—	—	—	-181	Keinle
2N11E12	497	WASC 2762	45404012123280 1	02N/11E-12BAB1	1772261	734539	678	—	—	—	678	—	443	—	—	233	—	33	Keinle
2N11E12	499	WASC 2757	—	—	1774669	731338	318	—	—	318	311	—	—	—	—	—	—	291	Keinle
2N11E12	1793	WASC 2755	—	—	1771418	733269	673	—	—	—	678	—	523	—	—	333	—	193	Keinle

Table A3. Elevation of the tops of hydrogeologic units in wells in the Mosier, Oregon, study area.—Continued

[This table is available as a digital spreadsheet at http://pubs.usgs.gov/sir/2012/5002. **State plane:** State plane Oregon North (North American Datum of 1927) northing and easting coordinates (Kienle, 1995; Jervey, 1996). **Land-surface elevation:** Determined using state plane northing and easting coordinates (Kienle, 1995; Jervey, 1996) and USGS 10 meter digital elevation model (DEM), North American Vertical Datum of 1988 (NAVD 88). **Elevations:** Determined using well log stratigraphic interpretations (Kienle, 1995; Jervey, 1996) and land-surface elevation, NAVD 88. Grey italicized elevations have been interpreted below the drilled well hole depths using cross-sectional maps (Jervey, 1996). **Abbreviations:** USGS, U.S. Geological Survey; –, no data]

Township range and section	County well No.	State well No.	USGS well No.	Station name	State plane Easting (feet)	State plane Northing (feet)	Land-surface elevation (feet)	Alluvium deposits (feet)	Glaciofluvial deposits (feet)	Chenoweth Formation (feet)	Pomona Member (feet)	Selah Interbed (feet)	Lolo Priest Rapids Member (feet)	Rosalia Priest Rapids Member (feet)	Quincy-Squaw Creek Interbed (feet)	Frenchman Springs Member (feet)	Grande Ronde Formation (feet)	Bottom elevation (feet)	Stratigraphic interpretation source
2N11E12	1794	WASC 2749	–	–	1771329	733553	668	–	–	–	668	–	450	–	280	260	–	95	Keinle
2N11E12	1974	WASC 1974	–	–	1771998	730645	303	–	–	–	303	–	163	–	–	–	–	113	Jervey
2N11E12	1984	WASC 1984	45402412233401	02.00N/11.00E-12BDB01	1771884	733062	645	–	–	645	250	–	125	–	–	-50	–	-100	Jervey
2N11E12	2030	WASC 2030	–	–	1775289	732153	389	–	389	–	254	130	70	–	–	-60	–	-110	Jervey
2N11E12	2092	WASC 2092	–	–	1774769	729773	382	–	–	382	372	–	192	–	–	–	–	142	Jervey
2N11E12	2112	WASC 2112	–	–	1770400	734250	607	–	–	–	607	442	422	–	280	227	–	75	Keinle
2N11E12	50012	WASC 50012	45402912225201	02.00N/11.00E-12ADB01	1774841	733278	317	–	317	–	202	37	-58	–	–	-128	–	-278	Jervey
2N11E13	506	WASC 2767	–	–	1775078	728456	374	–	–	374	364	–	214	–	–	–	–	164	Keinle
2N11E14	486	WASC 2769	45392812245001	02N/11E-14BCD1	1766555	727310	1,005	–	–	1,005	385	–	–	–	–	–	–	335	Keinle
2N11E14	508	WASC 2770	45391312245201	02N/11E-14CBD1	1766225	725976	1,026	–	–	1,026	570	–	–	–	–	–	–	520	Jervey
2N11E14	1879	WASC 1879	–	–	1766104	726889	959	–	–	959	545	–	–	–	–	–	–	495	Keinle
2N11E15	509	WASC 2772	45390012251901	02N/11E-15DDC1	1764254	724771	1,081	–	–	1,081	616	–	–	–	–	–	–	566	Jervey
2N11E23	510	WASC 2773	–	–	1768290	721270	892	–	–	892	442	–	–	–	–	–	–	392	Keinle
2N11E24	1795	WASC 2774	–	–	1775200	723800	493	–	–	–	–	–	493	–	–	433	–	343	Keinle
2N11E24	1976	WASC 2776	45385312231201	02N/11E-24ABC1	1773104	723546	516	–	–	–	516	491	391	–	311	–	–	271	Keinle
2N11E24	2029	WASC 2029	–	–	1773650	723240	505	–	–	–	505	434	428	–	330	–	–	280	Keinle
2N11E24	2032	WASC 2032	–	–	1772150	723050	669	–	–	–	669	–	529	–	–	–	–	479	Keinle
2N11E25	1869	WASC 1869	–	–	1771540	718470	1,123	–	1,123	–	–	–	–	–	–	–	–	645	Keinle
2N12E05	527	WASC 2820	45405612211201	02N/12E-05CCB1	1782322	735787	716	–	716	–	666	415	–	–	–	–	–	391	Keinle
2N12E05	535	WASC 2826	45410312111201	02N/12E-05CBC2	1781776	736704	807	–	–	–	807	–	650	–	380	–	–	259	Keinle
2N12E05	541	WASC 2815	45404612210501	02N/12E-05CCD1	1782313	735103	718	–	–	718	523	–	370	–	–	170	–	120	Keinle
2N12E05	543	WASC 50511	45405112203601	02.00N/12.00E-05DCB01	1784196	735343	691	–	691	–	662	–	–	–	–	–	–	566	Keinle
2N12E05	544	WASC 2805	–	–	1784684	734980	693	–	693	–	618	–	488	–	258	–	–	208	Jervey
2N12E05	545	WASC 2814	45405512203401	02N/12E-05DCB1	1784196	734925	698	–	698	–	603	–	488	–	268	–	–	218	Keinle
2N12E05	558	WASC 2843	45413212100001	02N/12E-05BBA1	1776342	739194	280	–	280	–	270	–	160	–	-40	-70	–	-350	Jervey
2N12E06	539	WASC 2823	–	–	1781091	738004	723	–	723	–	625	–	478	383	–	258	–	-107	Keinle
2N12E06	546	WASC 2838	45411012221401	02N/12E-06BCD1	1777430	737600	450	–	–	–	450	320	310	–	–	–	–	250	Keinle
2N12E06	547	WASC 2831	45412912222701	02N/12E-06BBD1	1775664	738435	312	–	312	–	279	–	–	–	–	–	–	186	Keinle
2N12E06	548	WASC 2834	–	–	1780445	738861	501	–	–	–	501	–	344	–	–	91	–	41	Jervey
2N12E06	549	WASC 2835	45405212222901	02N/12E-06CCB1	1776320	735470	435	–	435	–	–	–	–	–	–	–	–	160	Keinle

Table A3.   Elevation of the tops of hydrogeologic units in wells in the Mosier, Oregon, study area.—Continued

[This table is available as a digital spreadsheet at http://pubs.usgs.gov/sir/2012/5022. **State plane:** State plane Oregon North (North American Datum of 1927) northing and easting coordinates (Kienle, 1995; Jervey, 1996). **Land-surface elevation:** Determined using state plane northing and easting coordinates (Kienle, 1995; Jervey, 1996) and USGS 10 meter digital elevation model (DEM), North American Vertical Datum of 1988 (NAVD 88). **Elevations:** Determined using well log stratigraphic interpetations (Kienle, 1995; Jervey, 1996) and land-surface elevation, NAVD 88. Grey italicized elevations have been interpreted below the drilled well hole depths using cross-sectional maps (Jervey, 1996). **Abbreviations:** USGS, U.S. Geological Survey; –, no data]

Township range and section	County well No.	State well No.	USGS well No.	Station name	State plane Easting (feet)	State plane Northing (feet)	Land-surface elevation (feet)	Alluvium deposits (feet)	Glaciofluvial deposits (feet)	Chenoweth Formation (feet)	Pomona Member (feet)	Selah Interbed (feet)	Lolo Priest Rapids Member (feet)	Rosalia Priest Rapids Member (feet)	Quincy Squaw Creek Interbed (feet)	Frenchman Springs Member (feet)	Grande Ronde Formation (feet)	Bottom elevation (feet)	Stratigraphic interpretation source
2N12E06	550	WASC 2847	454127121223501	02N/12E-06BBC1	1779512	738514	393	–	393	–	379	–	203	–	–	13	–	-37	Keinle
2N12E06	551	WASC 2817	454056121211901	02N/12E-06DDA1	1781169	735951	686	–	686	–	600	436	427	–	–	230	–	180	Keinle
2N12E06	553	WASC 2830	–	–	1780093	739292	422	–	–	–	422	252	216	–	–	–	–	112	Keinle
2N12E06	554	WASC 2839	454115121215101	02N/12E-06ACC1	1779007	737986	413	–	413	–	388	–	208	–	–	28	–	-22	Keinle
2N12E06	555	WASC 2841	–	–	1780764	738725	545	–	545	–	525	–	375	–	–	175	–	125	Jervey
2N12E06	557	–	–	–	1777522	738236	354	–	354	–	340	–	200	–	–	1	–	-50	Jervey
2N12E06	561	WASC 2828	454051121221501	02.00N/12.00E-06CAD02	1777500	735550	482	–	482	–	137	–	–	–	–	–	–	99	Keinle
2N12E06	564	WASC 2833	–	–	1779744	739060	396	–	–	–	396	234	231	195	–	–	–	172	Keinle
2N12E06	565	WASC 2840	–	–	1781064	735553	682	–	–	682	385	–	–	–	–	–	–	527	Keinle
2N12E06	1759	WASC 2811	–	–	1781474	735908	690	–	690	620	450	–	350	–	–	160	–	110	Jervey
2N12E06	1921	WASC 1921	–	–	1776000	736347	554	–	–	–	554	356	311	–	–	102	–	95	Keinle
2N12E07	503	WASC 2763	453959121223701	02N/12E-07CCB1	1776050	730100	459	–	–	459	441	239	229	–	–	70	–	20	Keinle
2N12E07	567	WASC 2863	454003121223101	02N/12E-07CCA1	1776311	730847	482	–	–	482	369	192	162	–	–	22	–	-28	Keinle
2N12E07	568	WASC 2862	454004121211801	02.00N/12.00E-07DDA01	1781438	730934	682	–	–	682	579	–	430	–	–	214	–	202	Keinle
2N12E07	570	WASC 2860	454032121213101	02N/12E-07AAC2	1780366	733477	683	–	683	443	403	–	275	–	–	70	–	20	Keinle
2N12E07	571	WASC 2865	454003121215101	02N/12E-07DCB1	1779123	730816	413	–	413	–	385	–	273	–	–	73	–	23	Keinle
2N12E07	572	WASC 2870	454006121214501	02N/12E-07DBD1	1779634	730809	517	–	517	–	494	377	357	–	–	177	–	127	Keinle
2N12E07	573	WASC 2869	454032121213501	02N/12E-07AAC1	1779690	733478	659	–	659	520	375	–	235	–	–	40	–	-10	Jervey
2N12E07	574	WASC 2858	454043121223801	02.00N/12.00E-07BBB02	1775767	734747	341	–	341	–	–	–	135	–	–	76	–	6	Keinle
2N12E07	575	WASC 2861	454031121215701	02N/12E-07BDA1	1778739	733369	597	–	597	417	337	–	210	–	–	10	–	-40	Keinle
2N12E07	576	WASC 2866	454035121223101	02N/12E-07BBD1	1776342	733762	267	–	267	–	210	70	-40	–	–	–	–	-90	Keinle
2N12E07	577	WASC 2864	454010121224001	02.00N/12.00E-07CBC01	1775612	731416	400	–	400	–	330	165	139	–	–	-30	–	-80	Keinle
2N12E07	578	WASC 2867	454036121224001	02N/12E-07BBC1	1775835	734132	270	–	270	–	205	40	-60	–	–	–	–	-110	Keinle
2N12E07	581	WASC 2852	–	–	1781380	732290	562	–	–	562	515	339	331	–	–	125	–	75	Keinle
2N12E07	583	WASC 2857	453958121221701	02.00N/12.00E-07CCD01	1777330	730020	398	–	–	598	464	316	310	269	–	138	–	88	Keinle
2N12E07	584	WASC 2855	454008121215101	02.00N/12.00E-07DBC01	1779110	731100	486	–	486	–	472	336	294	–	–	166	–	116	Keinle
2N12E07	585	WASC 2856	–	–	1778129	731755	421	–	421	–	330	–	240	–	–	70	–	60	Jervey

Table A3. Elevation of the tops of hydrogeologic units in wells in the Mosier, Oregon, study area.—Continued

[This table is available as a digital spreadsheet at http://pubs.usgs.gov/sir/2012/5002. **State plane:** State plane Oregon North (North American Datum of 1927) northing and easting coordinates (Kienle, 1995; Jervey, 1996). **Land-surface elevation:** Determined using state plane northing and easting coordinates (Kienle, 1995; Jervey, 1996) and USGS 10 meter digital elevation model (DEM), North American Vertical Datum of 1988 (NAVD 88). **Elevations:** Determined using well log stratigraphic interpretations (Kienle, 1995; Jervey, 1996) and land-surface elevation, NAVD 88. Grey italicized elevations have been interpreted below the drilled well hole depths using cross-sectional maps (Jervey, 1996). **Abbreviations:** USGS, U.S. Geological Survey; –, no data]

Township range and section	County well No.	State well No.	USGS well No.	Station name	Easting (feet)	Northing (feet)	Land-surface elevation (feet)	Alluvium deposits (feet)	Glaciofluvial deposits (feet)	Chenoweth Formation (feet)	Pomona Member (feet)	Selah Interbed (feet)	Lolo Priest Rapids Member (feet)	Rosalia Priest Rapids Member (feet)	Quincy-Squaw Creek Interbed (feet)	Frenchman Springs Member (feet)	Grande Ronde Formation (feet)	Bottom elevation (feet)	Stratigraphic interpretation source
2N12E07	587	WASC 2872	454027121212501	02N/12E-07ADA1	1780967	733039	695	–	695	629	480	–	322	–	–	100	–	50	Keinle
2N12E07	589	WASC 2871	454020121223401	02N/12E-07BCC1	1776102	732495	443	–	–	443	285	208	–	–	–	38	–	-177	Keinle
2N12E07	1164	WASC 2875	454020121211901	02.00N/12.00E-07ADD01	1781370	732300	566	–	–	566	522	341	333	–	–	125	–	75	Keinle
2N12E07	1864	–	–	–	1779951	730309	529	–	–	529	500	–	310	–	–	130	–	80	Jervey
2N12E07	2090	WASC 2090	454032121215601	02.00N/12.00E-07BAD01	1778808	733541	593	–	593	412	352	168	153	105	–	8	–	-42	Keinle
2N12E07	2091	WASC 1905	454032121212001	02.00N/12.00E-07AAD01	1781200	733500	675	–	675	570	451	297	275	–	–	90	–	40	Keinle
2N12E08	579	WASC 2859	454023121210301	02.00N/12.00E-08BCD01	1782400	732590	686	–	–	686	546	–	376	–	–	176	–	126	Jervey
2N12E08	590	WASC 2877	–	–	1786505	733528	842	–	842	–	741	635	632	452	–	405	–	280	Keinle
2N12E08	591	WASC 2885	–	–	1786640	731670	1,015	–	–	1,015	852	–	750	–	–	530	–	415	Keinle
2N12E08	593	WASC 2886	–	–	1785440	733917	806	–	–	806	728	580	570	–	–	450	–	336	Keinle
2N12E08	594	WASC 2884	–	–	1785422	734632	712	–	–	712	682	–	–	–	–	–	–	625	Keinle
2N12E08	595	WASC 2887	–	–	1782017	734066	695	–	–	695	475	–	345	–	–	150	–	100	Keinle
2N12E08	597	WASC 2879	454037121205601	02.00N/12.00E-08BAC01	1783043	734051	743	–	–	743	556	–	415	–	–	–	–	337	Keinle
2N12E08	603	WASC 2890	454017121210001	02N/12E-08CBA1	1782713	731872	692	–	–	692	567	–	*400*	–	–	–	–	*350*	Keinle
2N12E08	604	WASC 2889	453956121210401	02N/12E-08CCD1	1781912	729620	790	–	–	790	*630*	–	*470*	–	–	*250*	–	*200*	Keinle
2N12E08	620	WASC 2906	–	–	1786540	732076	963	–	–	963	806	–	628	–	–	471	–	465	Keinle
2N12E08	1801	WASC 2873	–	–	1783650	729800	877	–	–	877	676	–	*520*	–	–	*320*	–	*270*	Keinle
2N12E08	1853	WASC 2888	454024121201401	02N/12E-08ADC1	1785800	733100	892	–	–	892	689	–	569	–	–	432	–	232	Keinle
2N12E08	1884	WASC 1884	453956121205501	02.00N/12.00E-08CDC01	1783200	729800	816	–	–	816	*675*	–	*525*	–	–	*305*	–	*255*	Keinle
2N12E08	2141	–	–	–	1786569	733067	875	–	–	875	795	–	655	–	465	450	–	240	Jervey
2N12E09	606	WASC 2899	–	–	1790339	733734	830	–	–	830	–	–	770	–	–	690	–	630	Jervey
2N12E09	607	WASC 2894	–	–	1787126	730021	1,153	–	–	1,153	850	–	760	–	–	*580*	–	*530*	Jervey
2N12E09	608	WASC 2891	–	–	1787725	732021	949	–	–	949	905	–	804	727	–	622	–	283	Keinle
2N12E09	609	WASC 2900	454041121185001	02N/12E-09AAA1	1791806	734257	698	–	–	–	–	–	698	–	–	568	–	488	Keinle
2N12E09	610	WASC 2907	–	–	1787340	732330	920	–	–	920	836	–	719	–	–	580	–	250	Keinle
2N12E09	611	WASC 2892	–	–	1788103	733517	834	–	–	834	749	–	680	–	–	–	–	492	Keinle

Table A3.   Elevation of the tops of hydrogeologic units in wells in the Mosier, Oregon, study area.—Continued

[This table is available as a digital spreadsheet at http://pubs.usgs.gov/sir/2012/5022. **State plane:** State plane Oregon North (North American Datum of 1927) northing and easting coordinates (Kienle, 1995; Jervey, 1996). **Land-surface elevation:** Determined using state plane northing and easting coordinates (Kienle, 1995; Jervey, 1996) and USGS 10 meter digital elevation model (DEM), North American Vertical Datum of 1988 (NAVD 88). **Elevations:** Determined using well log stratigraphic interpretations (Kienle, 1995; Jervey, 1996) and land-surface elevation, NAVD 88. Grey italicized elevations have been interpreted below the drilled well hole depths using cross-sectional maps (Jervey, 1996). **Abbreviations:** USGS, U.S. Geological Survey; –, no data]

Township range and section	County well No.	State well No.	USGS well No.	Station name	State plane Easting (feet)	State plane Northing (feet)	Land-surface elevation (feet)	Alluvium deposits (feet)	Glacio-fluvial deposits (feet)	Chenoweth Formation (feet)	Pomona Member (feet)	Selah Interbed (feet)	Lolo Priest Rapids Member (feet)	Rosalia Priest Rapids Member (feet)	Quincy Squaw Creek Interbed (feet)	Frenchman Springs Member (feet)	Grande Ronde Formation (feet)	Bottom elevation (feet)	Stratigraphic interpretation source
2N12E09	612	WASC 2898	45403212121200001	02.00N/12.00E-09BBC001	1786874	733580	839	–	–	839	761	656	–	–	–	–	–	652	Keinle
2N12E09	614	WASC 2897	–	–	1786874	733015	873	–	–	873	830	678	668	598	–	490	–	440	Keinle
2N12E09	615	WASC 2905	454024121192701	02N/12E-09BDA1	1789350	732730	851	–	–	851	835	–	736	–	–	640	–	230	Keinle
2N12E09	619	WASC 2896	–	–	1789400	732210	786	–	–	–	–	–	–	786	–	599	–	459	Keinle
2N12E09	621	WASC 2901	–	–	1788738	730426	1,047	–	–	1,047	977	859	856	–	–	–	–	847	Keinle
2N12E09	1901	–	–	–	1787078	730240	1,144	–	–	1,144	–	–	784	–	–	594	–	544	Jervey
2N12E09	2004	–	–	–	1790393	733202	800	–	–	800	–	–	750	–	–	660	–	30	Jervey
2N12E09	2136	WASC 2136	–	–	1786849	731135	1,081	–	–	1,081	–	–	776	–	–	560	–	510	Jervey
2N12E09	2148	–	–	–	1788499	732308	892	–	–	892	847	–	787	–	–	677	–	477	Jervey
2N12E09	2214	–	–	–	1789624	732471	769	–	–	–	–	–	769	–	–	625	–	590	Jervey
2N12E10	624	WASC 2909	–	–	1793504	732692	992	–	–	992	–	–	–	–	–	937	–	567	Jervey
2N12E10	625	WASC 1838	–	–	1792611	731832	970	–	–	970	–	–	–	–	–	925	–	565	Jervey
2N12E10	1838	WASC 2975	–	–	1792323	733815	702	–	–	–	–	–	702	–	–	620	–	480	Jervey
2N12E15	676	WASC 2971	–	–	1794438	724021	1,543	–	–	–	–	–	1,543	–	–	1,422	–	1,138	Keinle
2N12E15	677	WASC 3134	–	–	1792960	724940	1,374	–	–	–	–	–	1,374	–	–	1,339	–	1,066	Keinle
2N12E15	784	WASC 2970	453902121183801	02N/12E-15CCD1	1792865	724475	1,431	–	1,431	–	–	–	1,425	1,370	–	–	–	1,066	Keinle
2N12E15	1809	WASC 3106	–	–	1793500	724090	1,515	–	–	–	–	–	1,515	–	–	1,399	–	1,105	Keinle
2N12E16	679	WASC 3000	–	–	1788150	728904	1,151	–	–	1,151	929	–	821	–	–	–	–	808	Keinle
2N12E16	680	WASC 3002	453931121194701	02N/12E-16BCD1	1787189	727375	1,279	–	–	1,279	921	–	850	–	–	642	–	637	Keinle
2N12E16	681	–	453947121200101	02N/12E-16BBB1	1786729	728908	1,207	–	–	1,207	821	–	790	–	–	610	–	560	Keinle
2N12E16	682	WASC 3001	–	–	1790145	724866	1,309	–	–	1,309	–	–	1,114	–	–	924	500	450	Jervey
2N12E16	683	WASC 3004	453947121195101	02N/12E-16BBA1	1787595	728901	1,186	–	–	1,186	896	–	836	–	–	666	–	636	Keinle
2N12E16	685	WASC 3005	453943121195701	02N/12E-16BBC1	1787080	728543	1,242	–	–	1,242	844	–	790	–	–	617	–	480	Keinle
2N12E16	688	WASC 2989	453923121195501	02N/12E-16CBB1	1787513	726352	1,295	–	–	1,295	–	–	850	–	–	650	–	600	Keinle
2N12E16	689	WASC 2997	–	–	1787914	725233	1,322	–	–	1,322	–	–	–	–	–	–	–	1,252	Keinle
2N12E16	690	WASC 2982	–	–	1788103	724440	1,364	–	–	1,364	–	–	960	–	–	820	–	770	Jervey
2N12E16	692	WASC 2999	–	–	1787050	727050	1,288	–	–	1,288	–	–	853	–	–	682	–	598	Keinle
2N12E16	693	WASC 2987	453902121190901	02N/12E-16DCD1	1790420	724225	1,290	–	–	–	–	–	1,290	1,173	–	1,117	634	546	Keinle
2N12E16	694	WASC 2996	–	–	1789200	724900	1,329	–	–	1,329	1,148	–	1,063	–	–	759	–	709	Keinle
2N12E16	695	WASC 3003	–	–	1788074	726122	1,269	–	–	1,269	–	–	820	–	–	620	–	570	Jervey
2N12E16	697	WASC 2998	453913121194401	02N/12E-16CAC1	1788272	725616	1,272	–	–	1,272	–	–	857	–	–	637	–	587	Jervey
2N12E16	698	–	453904121192301	02N/12E-16DCC1	1789348	724698	1,351	–	–	1,351	–	–	1,049	951	–	867	–	715	Keinle

Table A3. Elevation of the tops of hydrogeologic units in wells in the Mosier, Oregon, study area.—Continued

[This table is available as a digital spreadsheet at http://pubs.usgs.gov/sir/2012/5002. **State plane:** State plane Oregon North (North American Datum of 1927) northing and easting coordinates (Kienle, 1995; Jervey, 1996). **Land-surface elevation:** Determined using state plane northing and easting coordinates (Kienle, 1995; Jervey, 1996) and USGS 10 meter digital elevation model (DEM), North American Vertical Datum of 1988 (NAVD 88). **Elevations:** Determined using well log stratigraphic interpretations (Kienle, 1995; Jervey, 1996) and land-surface elevation, NAVD 88. Grey italicized elevations have been interpreted below the drilled well hole depths using cross-sectional maps (Jervey, 1996). **Abbreviations:** USGS, U.S. Geological Survey; –, no data]

Township range and section	County well No.	State well No.	USGS well No.	Station name	State plane Easting (feet)	State plane Northing (feet)	Land-surface elevation (feet)	Alluvium deposits (feet)	Glaciofluvial deposits (feet)	Chenoweth Formation (feet)	Pomona Member (feet)	Selah Interbed (feet)	Lolo Priest Rapids Member (feet)	Rosalia Priest Rapids Member (feet)	Quincy Squaw Creek Interbed (feet)	Frenchman Springs Member (feet)	Grande Ronde Formation (feet)	Bottom elevation (feet)	Stratigraphic interpretation source
2N12E16	699	WASC 2988	—	—	1788481	724493	1,358	—	—	1,358	1,098	—	958	—	—	813	—	763	Keinle
2N12E16	702	WASC 3006	453930121193901	02N/12E-16BDC1	1788213	727202	1,209	—	—	1,209	—	—	1,005	—	—	700	—	604	Keinle
2N12E16	704	WASC 2981	—	—	1788452	726539	1,206	—	—	1,206	—	—	778	—	—	641	—	381	Keinle
2N12E16	1797	WASC 2977	—	—	1786840	725260	1,360	—	—	1,360	—	—	—	—	—	—	—	1,080	Keinle
2N12E16	1859	WASC 1859	—	—	1791238	725675	1,106	—	—	—	—	—	1,106	—	—	996	526	466	Jervey
2N12E16	1891	WASC 1891	—	—	1788850	726200	1,206	—	—	—	—	—	1,206	—	—	1,002	—	880	Keinle
2N12E16	1892	WASC 1892	—	—	1787800	728270	1,231	—	—	1,231	860	—	820	—	—	629	—	591	Keinle
2N12E16	1912	—	—	—	1788716	725399	1,290	—	—	1,290	—	—	985	—	—	700	—	650	Jervey
2N12E16	1978	WASC 1978	—	—	1790999	725364	1,128	—	—	—	—	—	1,128	—	—	988	488	438	Jervey
2N12E16	2020	WASC 2020	453940121191901	02.00N/12.00E-16ABC01	1789792	728141	956	—	—	—	—	—	956	—	—	836	—	756	Keinle
2N12E16	2044	WASC 2044	—	—	1789127	724495	1,352	—	—	1,352	—	—	1,087	—	—	870	—	820	Jervey
2N12E16	2144	WASC 2144	—	—	1788988	727892	1,078	—	—	1,078	—	—	780	—	—	600	—	550	Jervey
2N12E16	4108	—	—	—	1787732	727155	1,262	—	—	1,262	932	—	882	—	—	682	—	632	Jervey
2N12E17	705	WASC 3012	—	—	1785501	728771	1,166	—	—	1,166	—	—	—	—	—	—	—	1,106	Jervey
2N12E17	707	WASC 3010	453936121210901	02.00N/12.00E-17BCB01	1781900	727800	825	—	—	825	672	—	585	—	—	—	—	520	Keinle
2N12E17	708	WASC 3014	—	—	1783682	724338	1,230	—	—	1,230	850	—	772	712	—	—	—	695	Keinle
2N12E17	710	WASC 3013	—	—	1783730	724825	1,207	—	—	1,207	824	—	752	—	—	582	—	532	Keinle
2N12E17	1765	WASC 3008	—	—	1783700	727350	1,133	—	—	1,133	700	—	560	—	—	—	—	753	Keinle
2N12E17	1835	WASC 1835	—	—	1784580	729289	1,097	—	—	1,097	769	—	649	—	—	350	—	300	Jervey
2N12E17	1849	WASC 1849	—	—	1783647	727565	1,134	—	—	1,134	750	—	700	—	—	479	—	429	Jervey
2N12E17	1888	—	—	—	1786369	728659	1,171	—	—	1,171	690	—	570	—	—	540	—	490	Jervey
2N12E17	1997	WASC 1997	—	—	1783571	729076	940	—	—	940	715	625	606	—	—	400	—	350	Jervey
2N12E17	2010	WASC 2010	—	—	1783300	728500	1,092	—	—	1,092	805	—	745	—	—	454	—	404	Jervey
2N12E17	2015	WASC 2015	—	—	1783132	725483	1,145	—	—	1,145	802	—	732	—	—	555	—	505	Keinle
2N12E17	2017	WASC 2017	—	—	1782794	725012	1,072	—	—	1,072	755	—	645	—	—	542	—	492	Jervey
2N12E17	2019	WASC 2019	—	—	1782785	726510	895	—	—	895	713	—	613	—	—	465	—	415	Jervey
2N12E17	2037	—	—	—	1783174	727928	1,113	—	—	1,113	685	—	518	—	—	433	—	383	Jervey
2N12E17	2038	WASC 2038	—	—	1782600	729200	868	—	—	868	665	—	505	—	—	298	—	248	Keinle
2N12E17	2075	WASC 2075	453944121211301	02.00N/12.00E-17BBC01	1781700	728750	755	—	—	755	—	—	—	—	—	315	—	300	Keinle

Table A3. Elevation of the tops of hydrogeologic units in wells in the Mosier, Oregon, study area.—Continued

[This table is available as a digital spreadsheet at http://pubs.usgs.gov/sir/2012/5002. **State plane:** State plane Oregon North (North American Datum of 1927) northing and easting coordinates (Kienle, 1995; Jervey, 1996). **Land-surface elevation:** Determined using state plane northing and easting coordinates (Kienle, 1995; Jervey, 1996) and USGS 10 meter digital elevation model (DEM), North American Vertical Datum of 1988 (NAVD 88). **Elevations:** Determined using well log stratigraphic interpretations (Kienle, 1995; Jervey, 1996) and land-surface elevation, NAVD 88. Grey italicized elevations have been interpreted below the drilled well hole depths using cross-sectional maps (Jervey, 1996). **Abbreviations:** USGS, U.S. Geological Survey; –, no data]

Township range and section	County well No.	State well No.	USGS well No.	Station name	State plane Easting (feet)	State plane Northing (feet)	Land-surface elevation (feet)	Alluvium deposits (feet)	Glacio-fluvial deposits (feet)	Chenoweth Formation (feet)	Pomona Member (feet)	Selah Interbed (feet)	Lolo Priest Rapids Member (feet)	Rosalia Priest Rapids Member (feet)	Quincy Squaw Creek Interbed (feet)	Frenchman Springs Member (feet)	Grande Ronde Formation (feet)	Bottom elevation (feet)	Stratigraphic interpretation source
2N12E17	2142	WASC 2142	–	–	1783991	729310	926	–	–	926	696	–	540	–	–	330	–	280	Jervey
2N12E17	2158	–	–	–	1784547	724592	1,269	–	–	1,269	870	–	830	–	–	650	–	600	Jervey
2N12E17	2159	–	–	–	1786648	727692	1,256	–	–	1,256	–	–	861	–	–	681	–	631	Jervey
2N12E17	4144	–	–	–	1785188	729172	1,130	–	–	1,130	750	–	620	–	–	410	–	360	Jervey
2N12E18	713	WASC 3023	453921212131101	02N/12E-18DAB1	1780210	726420	692	–	–	692	622	–	512	–	332	322	–	212	Jervey
2N12E18	714	WASC 3017	–	–	1781040	725300	820	–	–	820	725	630	625	–	420	410	–	290	Keinle
2N12E18	716	WASC 3028	453944212121501	02N/12E-18BBD1	1777164	728445	643	–	–	643	520	–	368	–	–	–	–	183	Keinle
2N12E18	717	WASC 3029	453939121221201	02N/12E-18BDB1	1777541	728238	702	–	–	702	549	400	372	–	–	160	–	110	Keinle
2N12E18	718	WASC 3025	453931121220201	02N/12E-18BDD1	1778321	727545	757	–	–	757	592	427	407	–	207	187	–	177	Keinle
2N12E18	719	WASC 3024	453922121215801	02N/12E-18CAA1	1778350	726500	791	–	–	791	621	–	450	–	260	250	–	231	Keinle
2N12E18	721	WASC 3015	–	–	1778528	727764	719	–	–	719	629	–	479	–	–	209	–	159	Jervey
2N12E18	724	WASC 3026	453937121215801	02N/12E-18BDA1	1778581	727904	678	–	–	678	603	–	431	–	–	203	–	173	Keinle
2N12E18	725	WASC 3019	–	–	1780969	724476	801	–	–	801	761	662	653	–	–	435	–	386	Keinle
2N12E18	1820	WASC 3021	453949121220301	02N/12E-18BAB1	1778085	729231	665	–	–	665	506	320	300	–	–	120	–	112	Keinle
2N12E18	727	WASC 3036	453836121213201	02.00N/12.00E-19ADC01	1780431	721785	1,248	–	–	1,248	808	–	648	–	513	488	–	258	Keinle
2N12E18	729	WASC 3032	453811121212401	02N/12E-19DDD1	1780747	719219	1,372	–	–	1,372	991	–	870	–	–	730	270	220	Keinle
2N12E18	731	WASC 3034	–	–	1779601	722302	1,201	–	–	1,201	761	–	661	–	–	481	–	201	Jervey
2N12E19	733	WASC 3031	453825121221701	02.00N/12.00E-19CBD01	1777089	720521	513	–	513	–	–	–	–	469	–	421	–	367	Keinle
2N12E19	734	WASC 3040	453847121223701	02.00N/12.00E-19BCB01	1775725	723581	433	–	433	–	–	–	386	–	–	–	–	308	Keinle
2N12E19	1964	WASC 1964	–	–	1779911	720209	1,318	–	–	1,318	900	–	800	–	–	630	–	360	Jervey
2N12E19	2081	WASC 2081	–	–	1780160	720230	1,339	–	–	1,339	934	797	793	–	–	601	200	150	Keinle
2N12E19	–	WASC 3038	453859121222901	02N/12E-19BBB1	1776200	724200	413	–	413	–	–	–	385	–	–	–	–	323	Keinle
2N12E20	735	WASC 3050	–	–	1785881	723454	1,351	–	–	1,351	956	921	916	–	–	800	–	750	Keinle
2N12E20	736	WASC 3046	–	–	1783571	723636	1,233	–	–	1,233	865	–	813	–	–	643	–	593	Keinle
2N12E20	737	WASC 3045	–	–	1785900	721600	1,251	–	–	1,251	1,027	–	906	–	–	–	–	784	Keinle
2N12E20	738	WASC 3051	–	–	1784857	722430	1,244	–	–	1,244	824	–	764	–	–	604	–	554	Keinle
2N12E20	1515	–	–	–	1785803	722781	1,524	–	–	1,324	974	–	959	–	–	754	–	704	Jervey
2N12E20	1564	WASC 3043	–	–	1785460	724080	1,323	–	–	1,323	918	874	866	–	–	750	–	540	Keinle
2N12E20	1787	WASC 3042	–	–	1786444	723382	1,375	–	–	1,375	950	–	940	–	–	840	–	780	Jervey
2N12E20	1914	WASC 1914	–	–	1784419	723948	1,273	–	–	1,273	880	–	840	–	–	690	–	640	Jervey

Table A3. Elevation of the tops of hydrogeologic units in wells in the Mosier, Oregon, study area.—Continued

[This table is available as a digital spreadsheet at http://pubs.usgs.gov/sir/2012/5002. **State plane:** State plane Oregon North (North American Datum of 1927) northing and easting coordinates (Kienle, 1995; Jervey, 1996). **Land-surface elevation:** Determined using state plane northing and easting coordinates (Kienle, 1995; Jervey, 1996) and USGS 10 meter digital elevation model (DEM), North American Vertical Datum of 1988 (NAVD 88). **Elevations:** Determined using well log stratigraphic interpretations (Kienle, 1995; Jervey, 1996) and land-surface elevation, NAVD 88. Grey italicized elevations have been interpreted below the drilled well hole depths using cross-sectional maps (Jervey, 1996). **Abbreviations:** USGS, U.S. Geological Survey; –, no data]

Township range and section	County well No.	State well No.	USGS well No.	Station name	State plane Easting (feet)	Northing (feet)	Land-surface elevation (feet)	Alluvium deposits (feet)	Glaciofluvial deposits (feet)	Chenoweth Formation (feet)	Pomona Member (feet)	Selah Interbed (feet)	Lolo Priest Rapids Member (feet)	Rosalia Priest Rapids Member (feet)	Quincy Squaw Creek Interbed (feet)	Frenchman Springs Member (feet)	Grande Ronde Formation (feet)	Bottom elevation (feet)	Stratigraphic interpretation source
2N12E20	2069	WASC 2069	–	–	1784600	722600	1,249	–	–	1,249	859	–	789	–	–	639	–	589	Keinle
2N12E21	672	WASC 2976	–	–	1791700	721550	1,480	–	–	–	–	–	–	–	–	1,480	–	1,170	Keinle
2N12E21	701	WASC 3075	453858121192201	02N/12E-21ABB1	1789316	723863	1,385	–	–	1,385	–	–	1,130	–	–	930	–	586	Keinle
2N12E21	741	WASC 741	–	–	1788979	723226	1,442	–	–	1,442	–	–	1,082	–	–	972	–	922	Jervey
2N12E21	742	WASC 3074	–	–	1787806	723981	1,397	–	–	1,397	–	–	950	–	–	810	–	760	Jervey
2N12E21	743	WASC 3068	–	–	1789000	722840	1,439	–	–	1,439	1,180	–	1,150	–	–	1,020	–	970	Keinle
2N12E21	744	WASC 3059	–	–	1791581	722331	1,421	–	–	–	–	–	1,421	–	–	1,401	1,101	1,051	Jervey
2N12E21	745	WASC 3062	–	–	1790546	723863	1,278	–	–	–	–	–	1,278	–	–	1,153	–	1,151	Keinle
2N12E21	746	WASC 3066	–	–	1790850	721870	1,492	–	–	–	–	–	1,492	–	–	1,402	–	1,007	Keinle
2N12E21	747	WASC 3064	453845121191401	02.00N/12.00E-21ACA01	1789968	722651	1,360	–	–	–	–	–	1,360	–	1,240	1,220	–	1,130	Jervey
2N12E21	749	WASC 3070	–	–	1791318	722840	1,441	–	–	–	–	–	1,441	1,411	–	1,336	–	1,101	Keinle
2N12E21	753	WASC 3061	–	–	1790800	722200	1,454	–	–	–	–	–	1,454	–	1,380	1,351	–	914	Keinle
2N12E21	754	WASC 3060	453842121185801	02.00N/12.00E-21ADA01	1791191	722223	1,462	–	–	–	–	–	1,462	–	1,450	1,430	–	1,322	Keinle
2N12E21	757	WASC 3072	453835121192201	02N/12E-21ACC1	1789490	721591	1,423	–	–	–	–	–	1,423	–	–	1,293	–	1,272	Keinle
2N12E21	758	WASC 3055	453857121193001	02N/12E-21BAA1	1787851	723438	1,403	–	–	1,403	1,058	–	1,023	–	–	953	–	662	Keinle
2N12E21	759	WASC 3076	–	–	1790010	722021	1,412	–	–	–	–	–	1,412	–	–	1,283	–	742	Keinle
2N12E21	762	WASC 3054	–	–	1788500	720150	1,520	–	–	1,520	–	–	–	1,500	–	1,372	–	1,273	Keinle
2N12E21	764	WASC 3057	–	–	1788308	721501	1,361	–	–	1,361	–	–	1,266	1,248	–	1,238	–	1,199	Keinle
2N12E21	2042	WASC 2042	–	–	1789040	722840	1,439	–	–	1,439	1,195	1179	1,172	–	–	1,034	–	1,019	Keinle
2N12E21	2155	WASC 2155	–	–	1790000	721700	1,435	–	–	1,435	–	–	1,425	–	–	1,292	–	1,158	Keinle
2N12E21	4102	–	–	–	1790531	722387	1,422	–	–	–	–	–	1,422	–	1,352	1,307	–	1,157	Jervey
2N12E21	4103	–	–	–	1790407	722612	1,381	–	–	–	–	–	1,381	–	1,290	1,270	–	670	Jervey
2N12E22	765	WASC 3119	453849121180801	02N/12E-22ABC1	1794896	722997	1,621	–	–	–	–	–	–	–	–	1,621	1,200	1,150	Keinle
2N12E22	766	WASC 3115	453852121181001	02N/12E-22ABB1	1794690	723262	1,587	–	–	–	–	–	–	–	–	1,587	–	1,357	Keinle
2N12E22	767	WASC 3112	–	–	1795300	723730	1,652	–	–	–	–	–	–	1,652	–	1,602	1,072	1,047	Keinle
2N12E22	768	WASC 3122	453839121175501	02N/12E-22ACD1	1795700	721890	1,764	–	–	–	–	–	–	–	–	–	–	1,519	Keinle
2N12E22	771	WASC 3117	453841121175701	02N/12E-22ACA1	1795655	722258	1,741	–	–	–	–	–	–	–	–	1,741	1,261	1,211	Jervey
2N12E22	772	WASC 3118	–	–	1794788	722334	1,656	–	–	–	–	–	–	–	–	1,656	1,206	1,156	Keinle
2N12E22	776	WASC 3113	453825121181001	02N/12E-22DBC2	1794344	720808	1,665	–	–	–	–	–	–	–	–	1,665	1,340	1,335	Keinle
2N12E22	777	WASC 3131	453826121183901	02N/12E-22CBD1	1792629	720429	1,607	–	–	–	–	–	–	–	–	1,607	–	1,287	Keinle
2N12E22	779	WASC 3124	453840121183601	02N/12E-22BCA1	1792840	722220	1,471	–	–	–	–	–	–	–	–	1,471	–	1,073	Keinle

Table A3. Elevation of the tops of hydrogeologic units in wells in the Mosier, Oregon, study area.—Continued

[This table is available as a digital spreadsheet at http://pubs.usgs.gov/sir/2012/5002. **State plane:** State plane Oregon North (North American Datum of 1927) northing and easting coordinates (Kienle, 1995; Jervey, 1996). **Land-surface elevation:** Determined using state plane northing and easting coordinates (Kienle, 1995; Jervey, 1996) and USGS 10 meter digital elevation model (DEM), North American Vertical Datum of 1988 (NAVD 88). **Elevations:** Determined using well log stratigraphic interpretations (Kienle, 1995; Jervey, 1996) and land-surface elevation, NAVD 88. Grey italicized elevations have been interpreted below the drilled well hole depths using cross-sectional maps (Jervey, 1996). **Abbreviations:** USGS, U.S. Geological Survey; –, no data]

Township range and section	County well No.	State well No.	USGS well No.	Station name	State plane Easting (feet)	State plane Northing (feet)	Land-surface elevation (feet)	Alluvium deposits (feet)	Glaciofluvial deposits (feet)	Chenoweth Formation (feet)	Pomona Member (feet)	Selah Interbed (feet)	Lolo Priest Rapids Member (feet)	Rosalia Priest Rapids Member (feet)	Quincy-Squaw Creek Interbed (feet)	Frenchman Springs Member (feet)	Grande Ronde Formation (feet)	Bottom elevation (feet)	Stratigraphic interpretation source
2N12E22	783	WASC 3104	–	–	1793573	723638	1,550	–	–	–	–	–	1,550	–	1,490	1,460	1,160	1,120	Jervey
2N12E22	785	WASC 3133	453849121183401	02N/12E-22BBD1	1792925	722777	1,488	–	–	–	–	–	1,488	–	–	1,439	1,150	1,143	Keinle
2N12E22	787	WASC 3101	–	–	1792027	723533	1,325	–	–	–	–	–	1,325	–	–	1,245	–	855	Keinle
2N12E22	789	WASC 3081	–	–	1792400	723000	1,427	–	–	–	–	–	1,427	–	1,407	1,367	*1,117*	*1,067*	Keinle
2N12E22	790	WASC 3126	–	–	1792752	719518	1,712	–	–	–	–	–	–	–	–	1,712	1,450	1,450	Jervey
2N12E22	792	WASC 3095	–	–	1793637	720428	1,692	–	–	–	–	–	–	–	–	1,692	1,350	1,350	Jervey
2N12E22	800	WASC 3092	–	–	1797023	720021	1,812	–	–	–	–	–	–	–	–	1,812	*1,352*	*1,302*	Keinle
2N12E22	801	WASC 3102	–	–	1795406	719641	1,788	–	–	–	–	–	–	–	–	1,788	*1,368*	*1,318*	Jervey
2N12E22	804	WASC 3103	–	–	1794564	718880	1,791	–	–	–	–	–	–	–	–	1,791	–	1,491	Keinle
2N12E22	806	WASC 3099	–	–	1792642	722147	1,498	–	–	–	–	–	–	–	–	1,498	–	1,108	Keinle
2N12E22	807	WASC 3127	453835121184201	02N/12E-22BCC1	1792400	721560	1,603	–	–	–	–	–	1,603	–	–	1,574	1,000	903	Keinle
2N12E22	808	WASC 3109	–	–	1792250	721310	1,556	–	–	–	–	–	–	–	–	1,556	1,325	1,330	Keinle
2N12E22	809	WASC 3082	–	–	1793414	720450	1,682	–	–	–	–	–	–	–	–	1,682	1,340	1,312	Keinle
2N12E22	811	WASC 3129	–	–	1794300	720750	1,654	–	–	–	–	–	–	–	–	1,654	–	1,557	Keinle
2N12E22	812	WASC 3132	453828121183601	02N/12E-22CBA1	1792829	720829	1,586	–	–	–	–	–	–	–	–	1,586	*1,330*	*1,280*	Keinle
2N12E22	813	WASC 3107	–	–	1796865	720855	1,842	–	–	–	–	–	–	–	–	1,842	*1,332*	*1,282*	Keinle
2N12E22	814	WASC 3108	453816121175701	02N/12E-22DCA1	1795630	719708	1,800	–	–	–	–	–	–	–	–	1,800	–	1,520	Keinle
2N12E22	815	WASC 3138	453822121180401	02N/12E-22DBC1	1794750	720228	1,708	–	–	–	–	–	–	–	–	1,708	–	1,497	Keinle
2N12E22	817	WASC 3139	453812121174301	02N/12E-22DDD1	1796651	719332	1,766	–	–	–	–	–	–	–	–	1,766	*1,386*	*1,336*	Keinle
2N12E22	818	WASC 3084	453808121175701	02N/12E-22DCD1	1795622	718979	1,833	–	–	–	–	–	–	–	–	1,833	1,416	1,391	Keinle
2N12E22	820	WASC 3090	–	–	1796550	720150	1,800	–	–	–	–	–	–	–	–	1,800	1,255	1,157	Keinle
2N12E22	821	WASC 3091	–	–	1796570	719856	1,793	–	–	–	–	–	–	–	–	1,793	1,380	1,300	Jervey
2N12E22	822	WASC 3120	453834121180501	02N/12E-22ACC1	1795137	722035	1,708	–	–	–	–	–	–	–	–	1,708	*1,238*	*1,188*	Keinle
2N12E22	1757	WASC 3116	–	–	1794323	722907	1,556	–	–	–	–	–	–	–	–	1,556	*1,176*	*1,126*	Jervey
2N12E22	1817	WASC 3137	453841121181301	02.00N/12.00E-22BDA01	1794280	721910	1,642	–	–	–	–	–	–	–	–	1,642	–	1,372	Keinle
2N12E22	1842	WASC 1842	–	–	1794000	720750	1,542	–	–	–	–	–	–	–	–	1,642	*1,332*	*1,282*	Keinle
2N12E22	1909	–	–	–	1792154	719251	1,753	–	–	–	–	–	–	–	–	1,753	–	1,450	Jervey
2N12E22	2006	–	–	–	1796733	722405	1,736	–	–	–	–	–	–	–	–	1,736	–	1,321	Jervey
2N12E22	2016	–	–	–	1796833	723021	1,727	–	–	–	–	–	–	–	–	1,727	–	1,472	Jervey
2N12E22	2023	WASC 2023	–	–	1795912	722257	1,742	–	–	–	–	–	–	–	–	1,742	*1,274*	*1,224*	Keinle
2N12E22	2024	WASC 3123	453838121174801	02N/12E-22ADC1	1796430	721755	1,787	–	–	–	–	–	–	–	–	1,787	1,327	1,306	Keinle
2N12E22	2080	WASC 2080	–	–	1794130	723154	1,542	–	–	–	–	–	1,542	–	1,487	1,457	1,157	947	Jervey

Table A3. Elevation of the tops of hydrogeologic units in wells in the Mosier, Oregon, study area.—Continued

[This table is available as a digital spreadsheet at http://pubs.usgs.gov/sir/2012/5002. **State plane:** State plane Oregon North (North American Datum of 1927) northing and easting coordinates (Kienle, 1995; Jervey, 1996). **Land-surface elevation:** Determined using state plane northing and easting coordinates (Kienle, 1995; Jervey, 1996) and USGS 10 meter digital elevation model (DEM), North American Vertical Datum of 1988 (NAVD 88). **Elevations:** Determined using well log stratigraphic intepretations (Kienle, 1995; Jervey, 1996) and land-surface elevation, NAVD 88. Grey italicized elevations have been interpreted below the drilled well hole depths using cross-sectional maps (Jervey, 1996). **Abbreviations:** USGS, U.S. Geological Survey; –, no data]

Township range and section	County well No.	State well No.	USGS well No.	Station name	Easting (feet)	Northing (feet)	Land-surface elevation (feet)	Alluvium deposits (feet)	Glaciofluvial deposits (feet)	Chenoweth Formation (feet)	Pomona Member (feet)	Selah Interbed (feet)	Lolo Priest Rapids Member (feet)	Rosalia Priest Rapids Member (feet)	Quincy-Squaw Creek Interbed (feet)	Frenchman Springs Member (feet)	Grande Ronde Formation (feet)	Bottom elevation (feet)	Stratigraphic interpretation source
2N12E22	2108	WASC 2108	–	–	1794033	721105	1,615	–	–	–	–	–	–	–	–	1,615	–	1,390	Jervey
2N12E22	2182	–	–	–	1793518	719689	1,730	–	–	–	–	–	–	–	–	1,730	–	1,450	Jervey
2N12E23	824	–	–	–	1798003	723423	1,682	–	–	–	–	–	–	–	–	1,682	–	1,490	Jervey
2N12E23	826	WASC 3146	453838121171801	02N/12E-23BCD1	1798465	722068	1,814	–	–	–	–	–	–	–	–	1,814	–	1,414	Keinle
2N12E25	830	WASC 3153	453801121161701	02N/12E-25BBB1	1802778	718079	980	–	–	980	–	–	958	854	–	784	–	713	Keinle
2N12E25	831	WASC 3151	453805121155901	02N/12E-25BAB	1804061	718530	895	–	–	895	–	–	872	806	–	675	–	528	Keinle
2N12E26	837	WASC 3155	453800121163501	02N/12E-26AAB1	1801246	717628	1,125	–	–	1,125	–	–	1,106	1,037	–	–	–	1,025	Keinle
2N12E27	839	WASC 3159	453759121180901	02N/12E-27ABC1	1794645	717891	1,905	–	–	–	–	–	–	–	–	1,905	1,586	1,582	Keinle
2N12E27	840	WASC 3157	–	–	1796100	718600	1,841	–	–	–	–	–	–	–	–	1,841	1,461	1,451	Keinle
2N12E29	846	WASC 3161	–	–	1784378	716770	1,572	–	–	1,572	–	–	–	–	–	1,562	–	1,294	Keinle
2N12E29	1908	WASC 1908	–	–	1785660	717740	1,563	–	–	–	–	–	1,563	1,437	–	1,379	–	1,251	Keinle
2N12E29	1949	WASC 1949	–	–	1783400	718680	1,290	–	–	1,290	1,173	–	1,133	1,048	–	978	–	884	Keinle
2N12E30	728	WASC 3037	–	–	1780703	716570	1,508	–	–	1,508	–	–	1,313	1,243	–	1,163	693	438	Keinle
2N12E30	847	WASC 3164	453804121212001	02N/12E-30AAA1	1781400	718740	1,356	–	–	1,356	1,033	–	941	–	–	*780*	*340*	*290*	Jervey
2N12E30	848	WASC 3163	–	–	1780971	715366	1,653	–	–	1,653	–	–	1,560	–	–	1,380	*940*	*890*	Keinle
2N12E30	1302	WASC 3162	–	–	1779950	717800	1,386	–	–	1,386	1,091	–	1,020	–	–	*880*	*430*	*380*	Jervey
2N12E32	852	WASC 3168	–	–	1782717	713336	1,890	–	–	1,890	–	–	–	–	–	1,775	*1,295*	*1,245*	Jervey
2N12E32	853	WASC 3171	453655121203101	02N/12E-32ACC1	1784610	711310	2,151	–	–	–	–	–	–	–	–	2,151	*1,670*	*1,620*	Keinle
2N12E32	854	WASC 3169	–	–	1783497	712937	1,977	–	–	–	–	–	–	–	–	1,977	*1,472*	*1,422*	Jervey
2N12E32	855	WASC 3170	–	–	1783840	713400	2,037	–	–	–	–	–	–	–	–	2,037	1,437	1,177	Keinle

Appendix B. Estimation of Groundwater Recharge

B.1—Watershed Model

The Precipitation Runoff Modeling System (PRMS) watershed model (Leavesley and others, 1983; Leavesley and others, 1996) distributes daily precipitation over the land surface, uses daily air temperature to determine the rain/snow mix and evaporative losses, and partitions the remaining water through three interconnected subsurface reservoirs: the soil zone reservoir, subsurface reservoir, and the groundwater reservoir. Each reservoir drains at varying rates to the nearby stream with part of the groundwater reservoir also draining into the deeper groundwater system (groundwater sink) (fig. B1). In the watershed model, daily mean values are simulated for the storage in each reservoir, the rate of movement of water from one reservoir to the next, and the combined flow from the three reservoirs to the stream. Daily mean simulated streamflow was calibrated using observed daily mean streamflow from the Mosier Creek stream-gaging site (14113200, streamflow measurement site number 4, fig. 1). The groundwater sink represents water that drains from the groundwater reservoir that enters a regional aquifer or discharges to the stream downstream of the Mosier Creek gaging station, either to Mosier Creek or directly to the Columbia River. Groundwater recharge is the total amount of water entering the groundwater reservoir, and it equals the sum of the groundwater flow to nearby streams and the flow into the groundwater sink (fig. B1).

In this study, PRMS version 1.1.7 (Leavesley and others, 1983; Leavesley and others, 1996) was used to estimate recharge to the study area for 1955–2007. The model was developed and calibrated for the Mosier Creek gage basin, defined as the 41.5 mi² area upstream of the Mosier Creek gaging station for the period of available streamflow data from WY 1964–81, and 2006–07. Subsequently, the model area was expanded to include the entire basins of Mosier, Rock, and Rowena Creeks at their points of confluence with the Columbia River, and the simulation period was expanded to include available climate data so that recharge was estimated for all three basins for the entire period (1955–2007).

In PRMS, the model area is divided into smaller hydrologic response units (HRUs). Within each HRU, it is assumed that the hydrologic attributes controlling rainfall runoff and groundwater recharge are similar across the HRU. HRUs are delineated by the modeler in a manner that reflects spatially distributed attributes of elevation, slope, aspect, soils and land cover type. For the PRMS models created in this study, a combined total of 312 HRUs were delineated. The time-series data inputs to PRMS are daily total precipitation,

and daily maximum and minimum air temperature. Climate data were obtained from the Hood River climate site (National Weather Service (NWS) site number 354003 (Oregon Climate Service, 2009) (fig. 1). PRMS requires a complete climate data set, so occasional gaps were filled by interpolation or by regression with nearby sites.

Precipitation over the gage basin diminishes from west to east in a transition from the relatively wet part of the Western Cascades to the dry interior, and from the southern, upland part of the basin to the relatively low-elevation northern part of the basin near the Columbia River. Daily total precipitation at Hood River was distributed over each HRU based on the ratio of long-term (1971–2000) monthly average precipitation at the climate site and at each HRU. The average precipitation was derived from the Precipitation Elevation Regression on Independent Slopes Model (PRISM), which provides annual and monthly precipitation estimates over an 800 by 800 m grid of the State of Oregon (PRISM Group, 2010). The grid was intersected with the polygons representing the HRUs using ARC/INFO algorithms, resulting in a monthly average total precipitation at each HRU. Overall, precipitation over the gage basin was about 10 percent greater than at the climate site at Hood River, and precipitation at the HRUs varied from 50 to 200 percent of the value at Hood River. The derived ratio of monthly precipitation at the climate site to monthly precipitation at each HRU was multiplied by the measured daily precipitation at the climate site, resulting in precipitation at each HRU for each day during the simulation.

The general distribution of the PRISM-derived precipitation was tested at two precipitation measurement sites relatively close to the gage basin as a means to verify the ratio method for determining precipitation at each HRU. The average difference between the PRISM precipitation and measured precipitation at The Dalles (fig. 1), and at Crow Creek reservoir (approximately 2.5 mi south of the Mosier-Rock-Rowena Creek watershed) for the same period (1970–2000) was about 15 percent (Oregon Climate Service, 2009; Wasco County Extension Service, written commun., 2009).

Daily maximum and minimum air temperature at each HRU was based on the daily maximum and minimum air temperature at the climate site. Monthly lapse rates were applied to the difference in elevation between the climate site and each HRU. For PRMS, lapse rates are defined as the change in air temperature (degrees Fahrenheit) for every 1,000 ft. The lapse rates were predefined by analyzing air temperature records from the surrounding Mosier region, and then incorporated into PRMS as model parameter values.

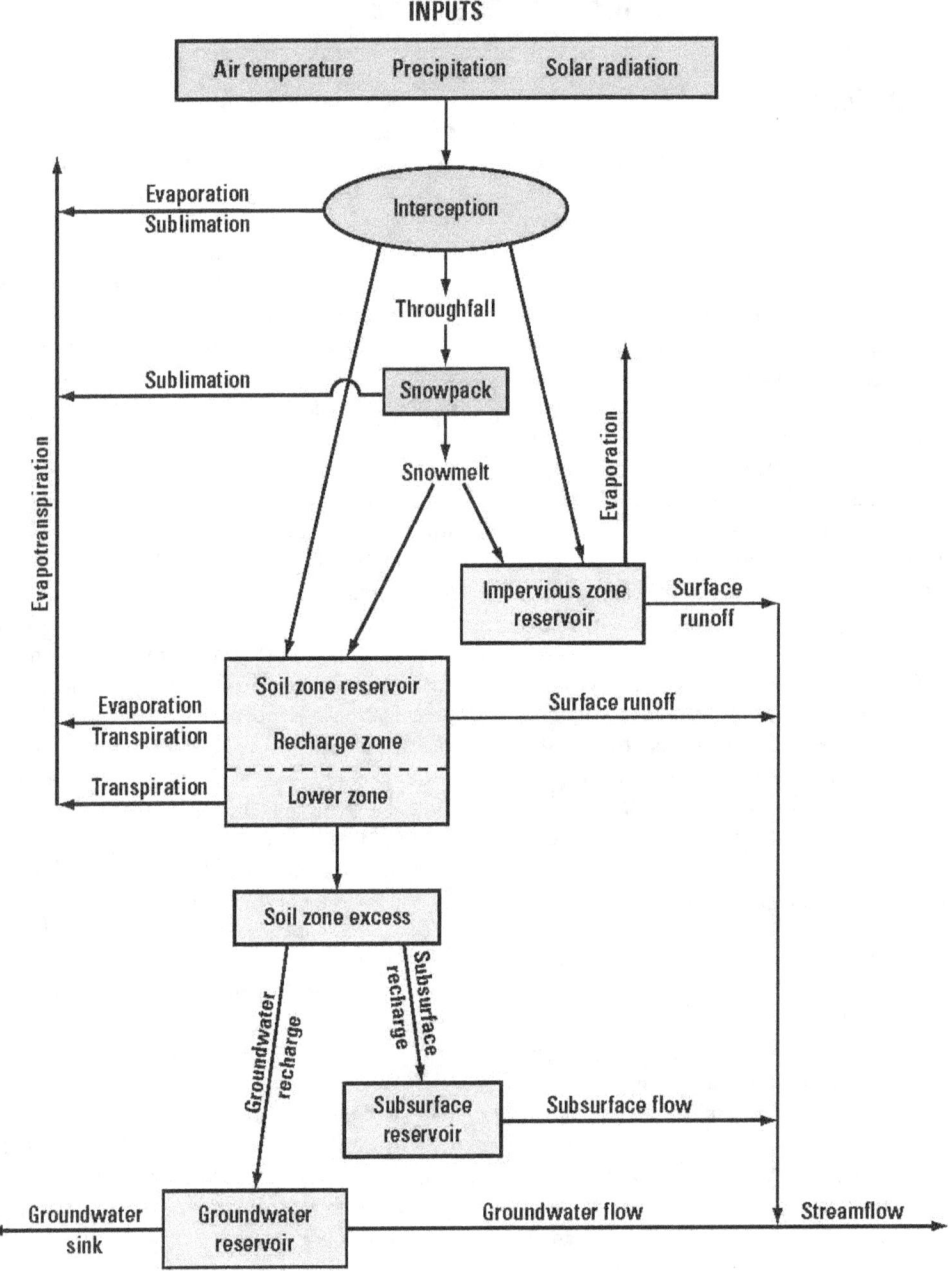

Figure B1.   The Precipitation Runoff Modeling System (PRMS). (Leavesley and others, 1996).

The simulation of daily mean streamflow derived from the PRMS model of the gage basin was verified by comparison with the observed daily mean streamflow hydrograph and by comparison of annual flow volume. The shape of the streamflow hydrograph and particularly the recession characteristics of streamflow were an indicator of model fit (Leavesley and others, 1996). The components of streamflow include relatively rapid surface runoff, attenuated subsurface flow where precipitation infiltrates and discharges to the stream—delayed and prolonged compared to the timing of surface runoff, and an even more delayed local groundwater flow component. A realistic balance between these three components results in a reasonable fit with the observed seasonal streamflow hydrograph. Many parameter values in PRMS were based on the underlying GIS layers derived from the GIS Weasel processing procedure. GIS Weasel is a software system designed to aid users in preparing spatial information as input to lumped and distributed parameter hydrologic simulation models (Viger and Leavesley, 2007). These parameters were not adjusted in model calibration due to inadequate physical-process data needed to justify that approach. Calibration of the model was accomplished by adjusting parameters within recommended bounds, and primarily included those controlling the rate of movement of water from the subsurface to the groundwater reservoir, from the subsurface and groundwater reservoirs to the stream, and from the groundwater reservoir to the groundwater sink.

The model was calibrated for general streamflow characteristics. As such, the model does not simulate individual storm events well. Observed streamflow increases and decreases more rapidly than the simulated streamflow. During the several-month-long dry period, simulated streamflow often is less than observed streamflow, indicating a dry stream during periods of measured low flow Figure B2 shows the ability of the model to simulate measured flows at the Mosier Creek gaging station for WYs 1973–77. This period was selected to represent a range of streamflow conditions of Mosier Creek. Total streamflows during WYs 1974 and 1975 were the highest and second highest during the simulation period. Alternatively, WYs 1977 and 1973 represented the lowest and second lowest total annual streamflows. A comparison of observed and simulated annual

flow volumes (fig. B3) was the basis for determining the PRMS groundwater sink parameter value. The groundwater sink parameter was manually adjusted iteratively, until the difference between simulated and observed annual flow volumes was minimized.

Following development of the PRMS model for the gage basin, the model extent was expanded to include the Mosier, Rock, and Rowena Creek basins. Although the Mosier and Rock Creek basins are not physically connected at a single outlet point (they each flow directly into the Columbia River), it was possible to define them within PRMS as a single watershed model due to their close proximity. Because there is no stream routing component in PRMS and the ordering of the HRUs does not matter, HRUs from both basins were included in the same model parameter file. A separate model of the Rowena Creek basin was prepared. The same method of HRU delineation used in the gage basin resulted in 133 and 70 HRUs for the Mosier/Rock basin and the Rowena basin, respectively. Due to lack of observed streamflow data for the mouth of Mosier Creek, Rock Creek, and Rowena Creek, the same set of parameters applied to the gage basin was used for the expanded model area. Identical methods were used to distribute the climate data over these basins (fig. B4). The model was initialized with WY 1953–54 climate data, and water-budget components were derived for the simulation period, WYs 1955–2007.

Average recharge in the Mosier, Rock, and Rowena Creek basins for the simulation period was 9.6 in., and generally follows the pattern of precipitation. The greatest recharge was in the upland area to the south, at about 19 in. Recharge diminished from the western part of the basin toward the east, where the lowest recharge was about 4 in. Of the total recharge, the local groundwater flow and sink components represented 43 and 57 percent of recharge, respectively. The groundwater-flow model extent is slightly larger than the area encompassed by the watershed models, so non-intersecting areas were estimated based on adjacent values from PRMS.

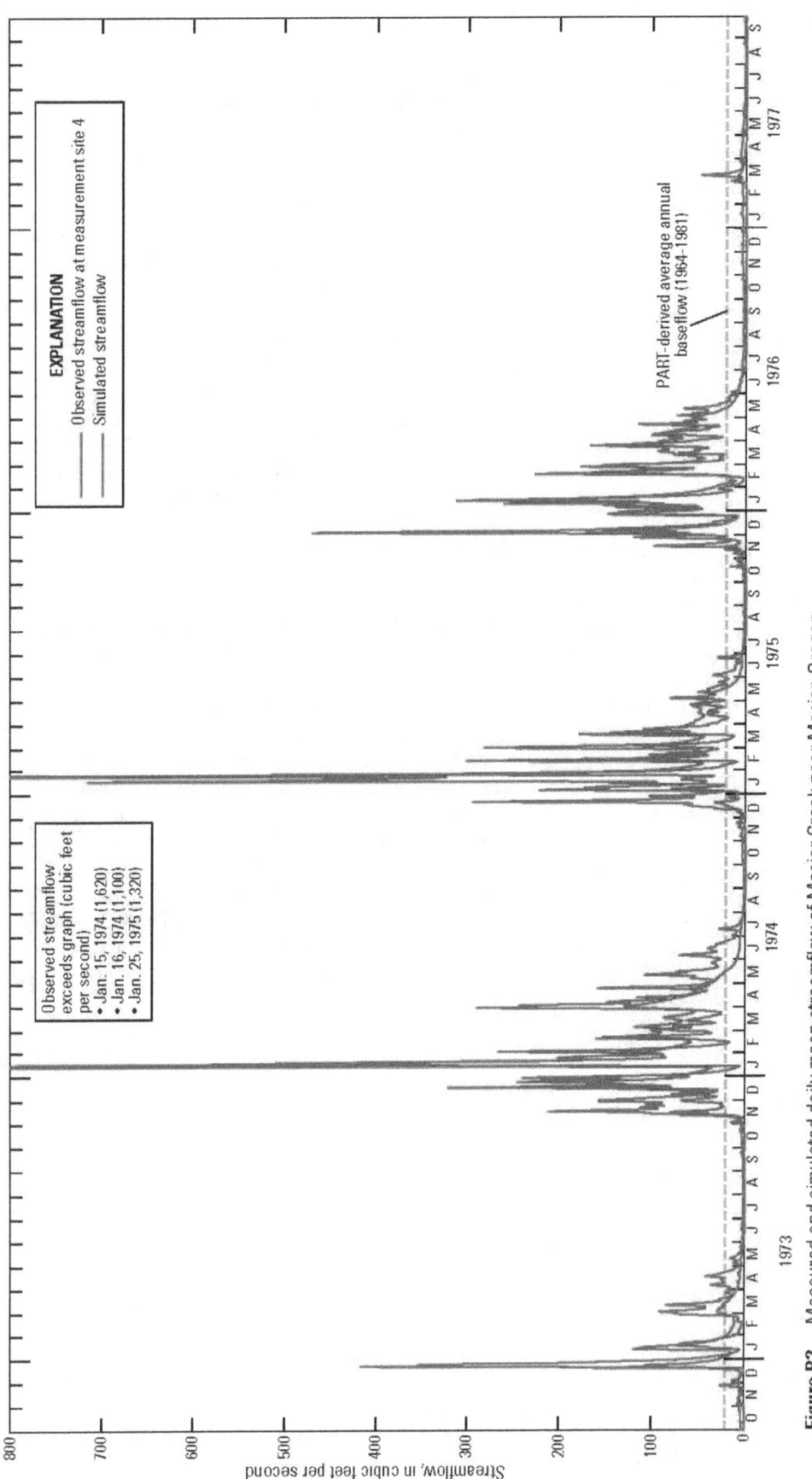

Figure B2. Measured and simulated daily mean streamflow of Mosier Creek near Mosier, Oregon.

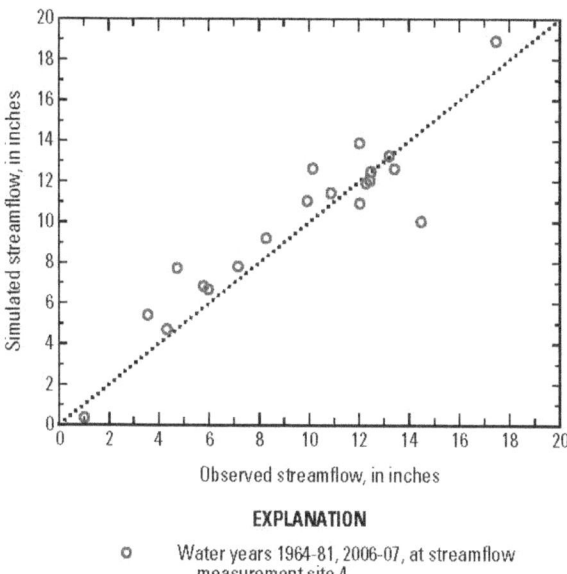

Figure B3. Measured and simulated annual mean streamflow of Mosier Creek near Mosier, Oregon.

B.2—Hydrograph Analysis

An independent estimate of recharge in the gage basin was provided by analysis of streamflow hydrographs using the programs RECESS and RORA (Rutledge, 1998). RECESS is a semi-automated procedure to determine the master recession curve (MRC) of streamflow recession. Using daily streamflow records from the Mosier Creek stream gage (14113200, streamflow measurement location number 4 [fig. 1]) the MRC was created using a manual iterative process. The final MRC was based on 23 periods of streamflow recession, beginning on the sixth day following a given peak, and extending for 20 days. RORA uses the recession-curve displacement method, incorporating the MRC to estimate recharge for each peak in the streamflow record, providing a daily estimate of recharge that is summed to annual values. The annual average recharge from RORA from 1964 to 1981 and 2006 to 2007 was 8.1 in., and varied from 1.0 to 14.3 in.

B.3—Comparison of Recharge Estimates

Recharge estimates from PRMS and RORA represent a range of values for the gage basin, and suggest a range of values for the study area. The two methods of estimating recharge are not strictly independent, as they both rely on the recession characteristics of streamflow. RORA assumes all groundwater movement is toward the stream, and in particular, that groundwater recharge emerges as groundwater discharge upstream from the streamflow site. The water balance of the PRMS model of the gage basin indicated that although part of the groundwater recharge (the local groundwater flow component) emerged upstream of the streamflow site, more than half the recharge (the groundwater sink component) emerged downstream of the streamflow site. The average local groundwater flow component from PRMS was 4.2 in., compared to 8.1 in. from RORA. By adding the groundwater sink component (from PRMS) of 5.5 in., recharge ranged from 9.7 to 13.6 in. from PRMS and RORA, respectively. The difference may be attributed to a fundamental difference in the definition of recharge. Although RORA derives recharge from each individual peak in streamflow, PRMS recharge is relatively conservative because it does not include subsurface flow from individual storms.

The PRMS-derived recharge values are of most use for the purposes of this study, because the pattern of recharge may be extended beyond the time period and spatial extent of the data available at the streamflow gaging station. Although the limited temporal and spatial extent of streamflow measurements precludes use of the RORA-derived recharge values directly, comparison of the range of values provided by the two independent methods provides a reasonable range over which to vary PRMS-derived recharge estimates during groundwater-flow simulation modeling.

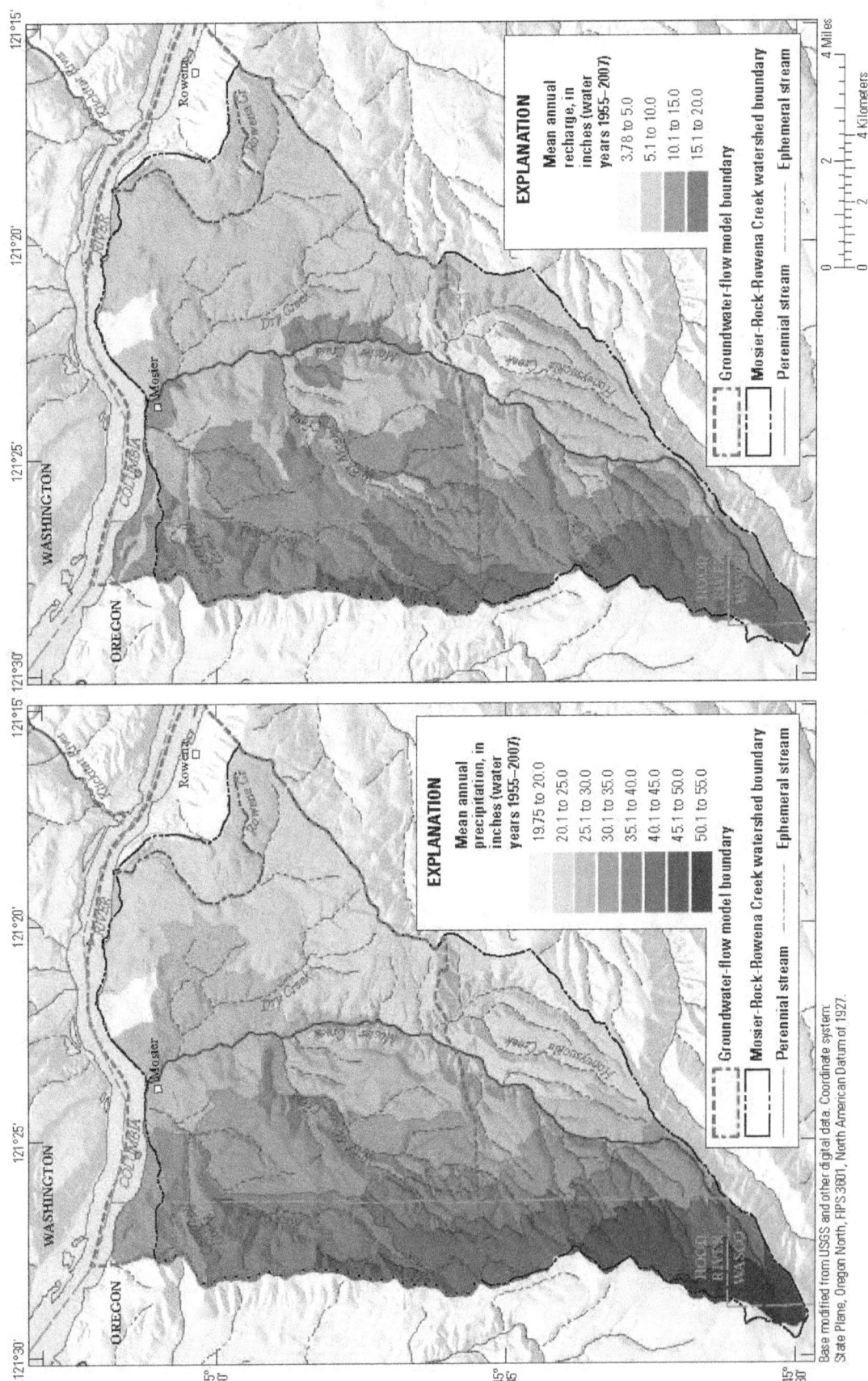

Figure B4. Model simulation results for precipitation and recharge in the Mosier, Rock, and Rowena Creek basins, Oregon, water years 1955–2007.

Appendix C. Estimation of Groundwater Fluxes to Mosier Watershed Streams

Groundwater discharge to Mosier Creek was identified by seepage studies and by base flow separation. Seepage is the exchange between groundwater and surface water at the streambed forming the boundary between a stream and an aquifer system. A seepage study is an indirect method of quantifying groundwater discharge (streamflow gains) or recharge (streamflow losses) at the streambed at numerous locations along the stream. A seepage study consists of a series of streamflow measurements made at numerous locations along a stream reach over a short period. After accounting for tributary inflows and streamflow diversions, the gain or loss in streamflow between one location and the next location downstream is attributed to interaction with the groundwater system. Base flow separation uses daily mean streamflow, and separates rapid runoff during storm events from the groundwater discharge component of streamflow. The base flow component of streamflow may be compared to recharge estimates by PRMS and RORA.

One finding of the recharge estimates from PRMS is the need to invoke the groundwater sink component, indicating that part of the groundwater recharge bypasses the gaging station site. Seepage studies can help identify the location where this groundwater discharges downstream of the stream gage.

In addition to streamflow measurements, water quality data were collected during some of the seepage studies. These data consisted of measurements of specific conductance at the time of streamflow measurement and continuous stream temperature data for several weeks surrounding the seepage study. An increase in specific conductance from one location to the next location downstream is an indication of groundwater discharge to the stream, owing to the relatively high (compared to that of stream water) specific conductance of groundwater in the Mosier basin. Similarly, during warm months, a decrease in stream temperature at subsequent sites downstream indicates discharge of relatively cool groundwater to the stream.

The low streamflow measured in summertime is base flow derived from groundwater discharge to the stream. Base flow separation is a semi-automated technique for separating the surface-runoff component of streamflow from the groundwater discharge component. It is based on daily mean streamflow at the Mosier Creek stream gage, and provides an annual estimate of the base flow component of streamflow at that location.

C.1—Seepage

Seepage studies of Mosier Creek were done in 1962 in a regional groundwater study (Newcomb, 1969), in 1986 as part of a water-availability study by the Oregon Water Resources Department (Lite and Grondin, 1988), and for the current study in 2005 and 2006. The 1962 seepage study extended far upstream of the current study, with limited detail in the current study area. The primary focus area of both the 1986 and the current study is the part of the basin from the confluence of West Fork Mosier Creek toward the mouth of Mosier Creek. In the current study, streamflow measurements were also made of Rock and Rowena Creeks. All measurement sites are listed in table C1 and their locations are shown on figure 1.

Two primary factors impose uncertainty in seepage studies—uncertainty in the streamflow measurements, and fluctuations in streamflow during the time of the study. The uncertainty of an individual streamflow measurement is affected by the uniformity of velocity, channel characteristics, and limitations of the meter in use. Most streamflow measurements made as part of this study were rated as "fair" using standard USGS qualitative rating methodology (Rantz, 1982), which assumes the streamflow is within 8 percent of the actual value. The accuracy of streamflow measurements of Mosier Creek was limited by the shallow depth of flow and low velocity. Considering the uncertainty associated with each streamflow measurement, the measured value represents a range of streamflow. If the magnitude of streamflow is large compared to the difference between streamflow at one location and the next location downstream, the net difference is often within the measurement uncertainty and therefore inconclusive. Second, accuracy of the seepage study is affected by temporal fluctuations in streamflow at each measurement location. At best, temporal flow fluctuation in a seepage study is known at a single location: the stream gage site. The gaging station was in operation during the seepage studies of July and September 2005, and in 2006. During summertime, and absent rainfall or withdrawals, streamflow is expected to be fairly steady; however, some natural diurnal fluctuations in streamflow do occur, typically caused by riparian evapotranspiration.

Table C1.   Streamflow and spring measurement sites in the Mosier, Oregon, study area.

[**Streamflow measurement site:** Refer to number in figure 1. **USGS site number:** Using this number, additional information is available from the USGS National Water Information System online at http://waterdata.usgs.gov/or/nwis. **Site name and location:** All sites near Mosier, Oregon, unless noted otherwise. **Abbreviations:** (a), daily mean streamflow; USGS, U.S. Geological Survey]

Streamflow measurement site	USGS site No.	Site name and location	River mile	Measurements made in years
1	453621121223200	Mosier Creek below Honeysuckle Creek	6.7	2005–06
2	453820121221500	Mosier Creek above Digger Road	4.1	1986, 2005–06
3	453853121223800	West Fork Mosier Creek at mouth		1962, 1986, 2005–06
4	14113200	Mosier Creek near Mosier	3.2	1962, 1963–81 (a), 2005–09 (a)
5	453922121223000	Mosier Creek at 1820 Mosier Creek Road	2.7	1986, 2005–06
6	453940121224200	Mosier Creek above Tanawasher Spring	2.1	1986, 2005–06
7	453951121224600	Mosier Creek below Tanawasher Spring	1.9	1986, 2005–06
8	454014121225200	Mosier Creek above dam	1.4	1986, 2005–06
9		Mosier Spring		1986, 2005–06
10	454041121230300	Mosier Creek above Dry Creek	0.9	1986, 2005–06
11	454042121230200	Dry Creek at mouth		4-05-05
12	454050121230600	Mosier Creek below Dry Creek	0.7	1962,1986, 2005–06
13	454105121233600	Mosier Creek at mouth between I-84 and highway 30	0.1	2006
14	454045121242800	Rock Creek near east tunnel portal		2005–06
15	454041121184800	Rowena Creek at Highway 30 near Rowena, Oregon		2005–06

A seepage study was done in September 1962 (table C2, fig. 11). Streamflow was measured at two locations coincident with the current study. Streamflow increased 0.4 ft³/s between the current (2009) stream gage site (streamflow measurement site number 4) and streamflow measurement site number 12, and of all seepage studies discussed in this report, represents the only gain during summertime in this reach. In addition, the magnitude of flow was greater than all other summertime streamflow measurements. These measurements were made prior to the installation of the gaging station, so it is unknown how representative these measurements are of low flow conditions, however weather conditions during the month prior to the 1962 measurements were seasonably warm and dry.

In 1986, seepage studies were done in June and August. The study done in June was disregarded due to uncertainty in methods. In August, streamflow measurements were made between site numbers 2 and 12 (table C2, figs. 1 and 11). Gains and losses in this reach from one measurement location to the next location downstream were as large as 0.5 ft³/s, greater than the measurement uncertainty. In the reach between site numbers 4 and 12, the loss in streamflow was about 0.1 ft³/s (10 percent), and was less than the measurement uncertainty of streamflow. The stream gage was not in operation during this study, however streamflow measurements at that site on two subsequent days indicated about a 50 percent fluctuation, suggesting caution regarding interpretations of gains and losses of similar magnitude during this study.

Seepage studies were made in April, July and September 2005, and May and August 2006, beginning at streamflow measurement site number 1. For consistency with previous studies, the upstream extent of the following analysis is at the stream gage site (streamflow measurement site number 4) (table C1), even though measurements were collected at locations upstream of the gaging station. Upstream seepage data were used during development of the conceptual model of groundwater flow and to aid in estimation of base flow flux calibration targets. The study of April 2005 was done prior to the re-installation of the gage.

During the July 2005 study, there were streamflow fluctuations owing to infiltration to the streambed and possibly to pumping from the stream. On the day of measurement, a tanker truck positioned just upstream of the Mosier Creek gaging station pumped from the creek four times for 15- to 30-minutes during the day. These withdrawals were evident in the streamflow record, where streamflow declined (and recovered) by about 50 percent each time. Translation of these pulses in streamflow may account for some of the fluctuation in measured streamflow at sites downstream. During the July 2005 study, no measurement was made at streamflow measurement site number 12 due to ponded conditions. The most downstream measurement location was the site upstream from the confluence with Dry Creek (streamflow measurement site number 10), and Dry Creek was dry during this study.

Table C2.   Streamflow measurements and seepage analysis, Mosier, Oregon, study area.

[**Streamflow measurement site:** Refer to number in <u>table C1</u>. **Measured streamflow:** Tributaries are in italics and underlined. **Gain (+) or loss (-) from next Mosier Creek measurement upstream:** Values in bold are greater than measurement uncertainty. **Abbreviations:** ft^3/s, cubic foot per second; (b), tributary treated as contribution, not a gain; (e), estimated]

Streamflow measurement site	Stream or spring	River mile	Measurement date and time	Measured streamflow (ft^3/s)	Measurement uncertainty (percent)	Gain (+) or loss (-) from next Mosier Creek measurement upstream (ft^3/s)	Gain (+) or loss (-) range of uncertainty from next Mosier Creek measurement upstream (ft^3/s)
4	Mosier Creek	3.2	09-12-62	2.84	5 (e)		
12	Mosier Creek	0.7	09-12-62	3.24	5 (e)	+0.40	+0.10 to +0.70
Summary		3.2 to 0.7	September 1962		5 (e)	+0.40	+0.10 to +0.70
2	Mosier Creek	4.1	08-19-86	1.30	8 (e)		
3	West Fork Mosier Creek (b)		08-19-86	*0.05*	8 (e)		
4	Mosier Creek	3.2	08-19-86	1.01	8 (e)	**-0.34**	-0.52 to -0.16
5	Mosier Creek	2.7	08-19-86	0.98	8 (e)	-0.03	-0.19 to 0.13
6	Mosier Creek	2.1	08-20-86	1.49	8 (e)	**0.51**	0.31 to 0.71
7	Mosier Creek	1.9	08-20-86	1.05	8 (e)	**-0.44**	-0.64 to -0.24
8	Mosier Creek	1.4	08-20-86	0.56	8 (e)	**-0.49**	-0.62 to -0.37
10	Mosier Creek	0.9	08-20-86	0.73	8 (e)	**0.17**	0.07 to 0.27
12	Mosier Creek	0.7	08-21-86	0.91	8 (e)	**0.18**	0.05 to 0.31
Summary		3.2 to 0.7	August 1986			-0.10	-0.25 to 0.05
1	Mosier Creek	6.7	04-05-05	13.6	8		
2	Mosier Creek	4.1	04-05-05	12.9	8	-0.7	-2.8 to 1.4
3	West Fork Mosier Creek (b)		04-05-05	*2.91*	8		
4	Mosier Creek	3.2	04-05-05	17.0	8	1.2	-1.4 to 3.8
5	Mosier Creek	2.7	04-05-05	16.6	5	-0.4	-2.6 to 1.8
6	Mosier Creek	2.1	04-05-05	16.9	8	0.3	-1.9 to 2.5
7	Mosier Creek	1.9	04-05-05	17.7	5	0.8	-1.4 to 3.0
8	Mosier Creek	1.4	04-05-05	16.9	5	-0.8	-2.5 to 0.9
9	Mosier Spring		04-05-05	0.10	5		
10	Mosier Creek	0.9	04-05-05	16.3	8	-0.6	-2.7 to 1.5
11	Dry Creek (b)		04-05-05	*0.57*	10		
12	Mosier Creek	0.7	04-05-05	17.7	5	0.8	-1.4 to 3.1
Summary		3.2 to 0.7	April 2005			0.1	-2.2 to 2.5
1	Mosier Creek	6.7	07-19-05	0.96	8		
2	Mosier Creek	4.1	07-19-05	1.10	8	0.14	-0.02 to 0.30
3	West Fork Mosier Creek (b)		07-19-05	*0.10*	10		
4	Mosier Creek	3.2	07-19-05	1.17	10	-0.03	-0.25 to 0.19
5	Mosier Creek	2.7	07-20-05	0.68	10	**-0.49**	-0.68 to -0.30
6	Mosier Creek	2.1	07-20-05	0.74	10	0.06	-0.08 to 0.20
7	Mosier Creek	1.9	07-20-05	1.07	10	**0.33**	0.15 to 0.51
8	Mosier Creek	1.4	07-20-05	0.41	5	**-0.66**	-0.79 to -0.53
9	Mosier Spring		07-20-05	0.00			
10	Mosier Creek	0.9	07-20-05	0.46	8	0.05	-0.01 to 0.11
11	Dry Creek (b)		07-20-05	0.00			
Summary		3.2 to 0.9	July 2005			-0.71	-0.87 to -0.55

Table C2.   Streamflow measurements and seepage analysis, Mosier, Oregon, study area.—Continued

[**Streamflow measurement site**: Refer to number in table C1. **Measured streamflow**: Tributaries are in italics and underlined. **Gain (+) or loss (–) from next Mosier Creek measurement upstream**: Values in bold are greater than measurement uncertainty. **Abbreviations**: ft³/s, cubic foot per second; **(b)**, tributary treated as contribution, not a gain; **(e)**, estimated]

Streamflow measurement site	Stream or spring	River mile	Measurement date and time	Measured streamflow (ft³/s)	Measurement uncertainty (percent)	Gain (+) or loss (–) from next Mosier Creek measurement upstream (ft³/s)	Gain (+) or loss (–) range of uncertainty from next Mosier Creek measurement upstream (ft³/s)
1	Mosier Creek	6.7	09-26-05	1.01	8		
2	Mosier Creek	4.1	09-26-05	1.01	10	0.00	-0.18 to 0.18
3	West Fork Mosier Creek (b)		09-27-05	*0.19*	10		
4	Mosier Creek	3.2	09-26-05	1.36	8	0.16	-0.07 to 0.39
5	Mosier Creek	2.7	09-26-05	1.08	8	-0.28	-0.48 to 0.08
6	Mosier Creek	2.1	09-26-05	1.03	10	-0.05	-0.24 to 0.14
7	Mosier Creek	1.9	09-26-05	1.07	8	0.04	-0.15 to 0.23
8	Mosier Creek	1.4	09-26-05	1.01	5	-0.06	-0.20 to 0.08
9	Mosier Spring		09-26-05	0.05	5		
10	Mosier Creek	0.9	09-26-05	1.13	5	0.12	0.01 to 0.23
11	Dry Creek (b)		09-26-05	*0.00*			
12	Mosier Creek	0.7	09-26-05	1.02	10	-0.11	-0.27 to 0.05
Summary		**3.2 to 0.7**	**September 2005**			-0.34	-0.55 to -0.13
1	Mosier Creek	6.7	05-16-06	6.05	8		
2	Mosier Creek	4.1	05-16-06	6.90	8	0.85	-0.19 to 1.89
3	West Fork Mosier Creek (b)		05-15-06	*0.94*	8		
4	Mosier Creek	3.2	05-16-06	7.66	8	-0.18	-1.42 to 1.06
5	Mosier Creek	2.7	05-16-06	8.56	5	0.90	-.14 to 1.94
6	Mosier Creek	2.1	05-16-06	7.95	5	-0.61	-1.44 to 0.22
7	Mosier Creek	1.9	05-16-06	7.32	8	-0.63	-1.61 to 0.35
8	Mosier Creek	1.4	05-16-06	7.31	5	-0.01	-0.96 to 0.94
9	Mosier Spring		05-15-06	0.08	5		
10	Mosier Creek	0.9	05-16-06	8.56	8	1.25	0.20 to 2.30
11	Dry Creek (b)		05-15-06	*0.33*	10		
12	Mosier Creek	0.7	05-16-06	8.50	8	-0.39	-1.78 to 1.00
13	Mosier Creek	0.1	05-16-06	8.81	5	0.31	-0.81 to 1.43
Summary		**3.2 to 0.7**	**May 2006**			0.51	-0.81 to 1.83
1	Mosier Creek	6.7	08-01-06	1.38	8		
2	Mosier Creek	4.1	08-01-06	1.19	10	-0.19	-0.42 to 0.04
3	West Fork Mosier Creek (b)		08-01-06	*0.05*	10		
4	Mosier Creek	3.2	08-01-06	1.26	8	0.02	-0.21 to 0.25
5	Mosier Creek	2.7	08-01-06	0.93	8	-0.33	-0.51 to -0.15
6	Mosier Creek	2.1	08-01-06	1.12	10	0.19	0.00 to 0.38
7	Mosier Creek	1.9	08-01-06	0.92	8	-0.20	-0.39 to -0.01
8	Mosier Creek	1.4	08-01-06	0.66	8	-0.26	-0.39 to -0.13
9	Mosier Spring		07-31-06	0.00			
10	Mosier Creek	0.9	08-01-06	0.59	10	-0.07	-0.18 to 0.04
11	Dry Creek (b)		08-01-06	*0.00*			
12	Mosier Creek	0.7	08-01-06	0.79	10	0.20	0.06 to 0.34
13	Mosier Creek	0.1	08-10-06	0.59	8	-0.20	-0.33 to -0.07
Summary		**3.2 to 0.7**	**August 2006**			-0.47	-0.65 to -0.29

Streamflow decreased 0.71 ft³/s between streamflow measurement sites 4 and 10, and this decrease was greater than the measurement uncertainty. Despite the pumping, losses of about 0.5 ft³/s between streamflow measurement site numbers 8 and 10 are considered accurate owing to the consistently low streamflow measured at these sites about 2 mi downstream of the location of withdrawal.

The streamflow measurements of September 2005 and August 2006 were made during relatively stable, low streamflow. The loss observed between streamflow measurement site numbers 4 and 12 in September 2005 (–0.34 ft³/s) and August 2006 (–0.47 ft³/s) were both greater than the measurement uncertainty.

Although the 2005 and 2006 seepage studies indicated net losses over the length of the study reach, changes in specific conductance and continuous stream-temperature data measured upstream and downstream of Mosier Spring (streamflow site 9) indicated some groundwater inflow. However, the groundwater inflow was not of a sufficient magnitude to be detected in the streamflow measurements. Specific conductance measurements (fig. 11) indicated generally similar values at sites upstream of streamflow measurement site number 8 and increases at streamflow measurement site numbers 10 and 12. The specific conductance of springs, seeps and Dry Creek was measured, and ranged from two to three times the value of Mosier Creek. Of particular interest is the relatively sharp increase in specific conductance in late summer of 2005 and 2006 between streamflow measurement site numbers 10 and 12, encompassing the tributary Dry Creek, which was dry during these times. Although streamflow measurements at these sites indicated a slight loss in 2005 and a slight gain in 2006, specific conductance increased sharply in both years. The only decrease in specific conductance from one location to the next location downstream was in August 2006, between streamflow measurement site number 12 and streamflow measurement site number 13. Streamflow measurement site number 13 is located at the mouth of Mosier Creek, near the elevation of the Columbia River. The decrease in specific conductance at this site suggests interaction with the Columbia River. Stream temperature was another indicator of interaction of the stream and the surrounding aquifer. At sites upstream, stream temperature gradually increased at each subsequent location downstream. Between streamflow measurement site numbers 8 and 10, in both late summer 2005 and 2006, stream temperature decreased between 1 and 2 °C, indicating groundwater contributions to the creek.

In addition to seepage studies of Mosier Creek, measurements of flow in other streams in the study area were made and used to verify flow simulated by the PRMS models for those locations. These consisted of a single streamflow measurement of both Rowena and Rock Creeks in 2005 and 2006. Streamflow of Rowena Creek at Highway 30 (streamflow site 15) in April 2005 and May 2006 was 0.08 and 0.18 ft³/s, respectively, and was zero (dry) during the summer seepage studies. Streamflow of Rock Creek (streamflow measurement site number 14) was measured upstream of a large quarry, and the creek was flowing during all seepage studies. Streamflow in 2005 (the average of the July and September measurements) was about 0.10 ft³/s, and was 0.05 ft³/s in July 2006.

Comparing the September 1962 streamflow measurements to later measurements, there is an apparent reduction in total base flow and a shift from net gaining to losing in the reach between the stream gage site and the Rocky Prairie thrust fault (fig. 11). Although these patterns of base flow are the expected result of groundwater declines to the south of the Rocky Prairie thrust fault, precipitation at the proximal Hood River rain gage was significantly higher during August and September of 1962 than for the periods preceding all other measurements, obfuscating clear linkages between the declining groundwater levels and the magnitude of base flow reduction in this area.

C.2—Base Flow Separation

The base flow component of stream flow was determined using the program PART using default parameter settings (Rutledge, 1998). PART uses streamflow partitioning to estimate a daily record of base flow from the stream flow hydrograph. The method assumes base flow equals streamflow on successive days when the streamflow is slowly receding, and linearly interpolates base flow for other days. Applied to multi-year periods, base flow separation provides an estimate of groundwater discharge. Expressed in inches, annual base flow totals were computed by summing monthly base flow totals by water year (October 1 to September 30). The lowest (1.0 in. in 1977) and highest (13.3 in. in 1974) annual totals were coincident with the lowest and highest occurrences of annual precipitation and streamflow for the period of record (WYs 1964–81 and 2006–07). Mean annual base flow for the same period was 6.9 in. (21.1 ft³/s), or about 70 percent of stream flow. During low-flow years almost the entire stream flow for the year was base flow. During summertime (July through September) base flow was 0.14 in., (1.7 ft³/s) or about 95 percent of streamflow.

Appendix D. Estimation of Pumping

Of the 19 high-capacity irrigation wells pumped during 2006 (fig. 14), 14 wells were equipped with flow meters, and pumping was estimated for the remaining 5 wells (see Pumping of Groundwater). These data and agricultural records were examined to estimate pumping at each well for 1966–2006 (fig. 13).

The USGS deployed inline turbine-type flow meters on 12 irrigation wells to measure irrigation water use for the 2006 irrigation season (fig. 14). The irrigation season occurs from April to September. The 12 flow meters, plus 2 owner-installed flow meters, measured an estimated 88 percent of the total irrigation water use (table D1). The 14 measured wells accounted for 74 percent of irrigation wells in the basin. Total measured discharge was 652 acre-ft and for individual wells ranged from 4 to 120 acre-ft with the average being 47 acre-ft. This water was applied to 741 acres, 86 percent of groundwater-irrigated acres in the Mosier basin. The application rate ranged from 0.14 to 1.86 ft/yr, with the average being 0.88 ft/yr.

To estimate 2006 water use for the five wells without an installed flow meter, coefficients were calculated from wells with flow meters (table D1), using the measured water applied, number of acres irrigated, and the type of irrigation system. The coefficients apply to the dominant crop (cherry orchards). The wells to be estimated had only two configurations of irrigation systems: Drip and micro spray used in conjunction or micro spray only. For these two irrigation methods, coefficients were developed:

Drip and micro-spray irrigation used in conjunction = 0.85 ft of water applied/acre.

Micro-spray only = 1.00 ft of water applied/acre.

These coefficients were used to estimate the remaining 12 percent of the total irrigation water use. Estimated well discharges ranged from 5 to 29 acre-ft with the average being 17 acre-ft. Aggregate unmetered water use for the 2006 irrigation season totaled 86 acre-ft. This water was applied to 118 acres, 14 percent of groundwater-irrigated acres in the Mosier Basin (table D1).

Table D1.   Irrigation well pumping in the Mosier, Oregon, study area, 2006.

[Alternating shading shows relationship between wells pumped and the associated areas irrigated. **Abbreviations**: USGS, U.S. Geological Survey]

Farm	Site identification No.	Station name	Measurement method	2006 groundwater pumped (acre-feet)	2006 area irrigated on each farm (acres)	2006 application rate by farm (acre-feet per year)
A	454031121224001	02N/11E-12AAD1	Estimated	5.4	5.4	1.00
B	454029121225201	02.00N/11.00E-12ADB01	Owner flowmeter	3.7	27.0	0.14
C	454013121225902	02.00N/11.00E-12DAB02	USGS flowmeter	79.0	79.0	1.00
D	453943121224901	02.00N/11.00E-13AAD01	USGS flowmeter	57.0	82.0	0.70
E	454055121203401	02N/12E-05DCB1	Estimated	17.8	21.0	0.85
F	454057121220301	02.00N/12.00E-06CAD01	Owner flowmeter	12.8	15.0	0.85
A	454052121223301	02.00N/12.00E-06CCB03	Estimated	28.6	33.6	0.85
G	454032121213501	02N/12E-07AAC1	USGS flowmeter	60.6		
	454032121215601	02.00N/12.00E-07BAD01	USGS flowmeter	67.5		
			Farm total:	128.0	134.0	0.96
H	454032121213101	02N/12E-07AAC2	USGS flowmeter	37.2		
	454032121212001	02.00N/12.00E-07AAD01	USGS flowmeter	21.4		
			Farm total:	58.6	78.0	0.75
I	454020121211901	02.00N/12.00E-07ADD01	USGS flowmeter	14.2	27.0	0.53
J	454020121223401	02N/12E-07BCC1	USGS flowmeter	35.5	60.0	0.59
K	454011121223901	02.00N/12.00E-07CBB01	Estimated	17.0	40.8	0.42
L	454008121215101	02.00N/12.00E-07DBC01	USGS flowmeter	46.0	24.7	1.86
M	454004121211801	02.00N/12.00E-07DDA01	USGS flowmeter	53.0	39.4	1.35
K	453949121220301	02N/12E-18BAB1	USGS flowmeter	119.6		
	453942121221501	02N/12E-18BBD1	USGS flowmeter	44.9		
			Farm total:	164.5	174.0	0.95
N	453921121213101	02N/12E-18DAB1	Estimated	17.3	17.3	1.00
			Total:	738.5	858.2	

For 1966–2006, annual irrigation water use was assumed constant for each fully established farm (except for minor fluctuations as reported in owner accounts). Increases in total irrigation pumpage during this period (fig. 13) correspond to the establishment of new farms. This assumption is supported by the water use estimates of Lite and Grondin (1988) and those collected by USGS in 2006. Lite and Grondin determined irrigation water use for 1986 was 570 acre-ft applied to 550 acres, and USGS estimated that for 2006, 566 acre-ft of water was applied to 621 acres on the same farms. The increase in acreage supplied by the nearly equivalent pumpage corresponds to improvements in irrigation and other agricultural practices that have resulted in increased tree density and lower water use per tree (fig. D1). Irrigated acreage for 2006 was estimated using the Oregon Water Resource Department's Water Rights Information System (WRIS) (Oregon Water Resources Department, 2006), Farm Service Agency Common Land Unit (CLU) GIS maps (data provided by James Bishop, County Executive Director, Farm Service Agency, U.S. Department of Agriculture, written commun., 2006), aerial photography, and owner accounts. The

fraction of acreage irrigated by micro spray, impact sprinkler, and drip irrigation methods was determined from discussions with owners and site visits.

Water use by tree estimates for mature cherry trees in nearby The Dalles, Oregon, are 1,250–2,500 gal per tree per yr for drip irrigation, 4,300 gallons per tree per year for micro spray irrigation, and 6,000 gallons per tree per year for impact sprinkler irrigation (J.P. le Roux, , IRRINET LLC, written commun., 2008). These rates compared favorably with Mosier per tree rates computed by dividing the total estimated irrigation pumping by the estimated total acreage irrigated and the estimated average cherry trees per acre for Wasco County (U.S. Department of Agriculture, 2006). Rates ranged from an average of about 5,000 gallons per tree per year in 1986 to 1,800 gallons per tree per year in 2006, which corresponds to the historical shift from less efficient to more efficient irrigation methods. Because USDA average tree density estimates were only available for 1986, 1993, and 2006, figure D1 was constructed by linearly interpolating tree density between these periods.

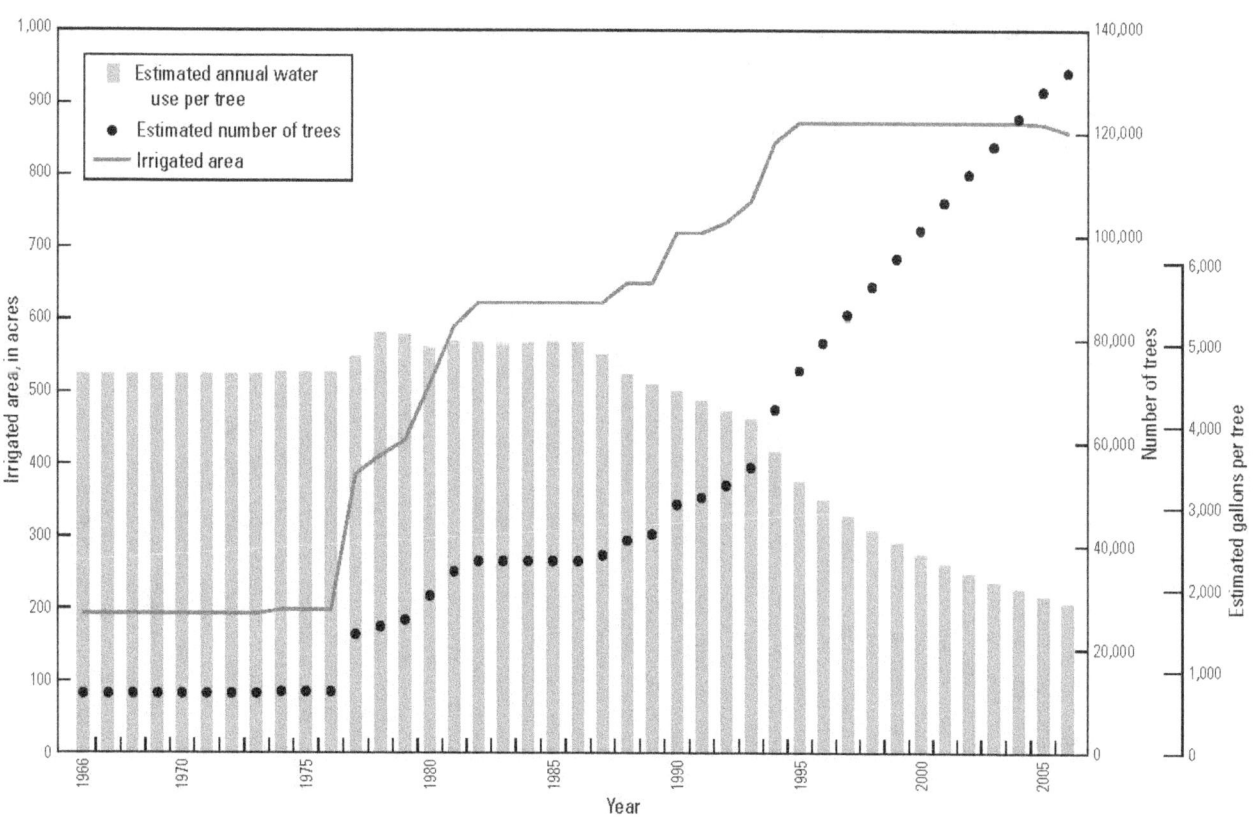

Figure D1.   Irrigated acreage, estimated number of fruit trees, and estimated consumptive use per tree, 1966–2006, in the Mosier, Oregon, study area.

The city of Mosier provided the USGS with meter readings for the primary public-supply well and water-use estimates for the backup public-supply well for 1989–2006. In 2006, the combined pumpage was 90 acre-ft (figs. 13 and 14). Prior to 1989 public-supply water use was estimated by the USGS based on 1989 water use and population.

Self-supplied domestic water use was estimated based on a number of assumptions about the population and per capita water-use behavior. Residences were identified using tax lot data. Additional residences were identified where known locations of self-supplied domestic wells existed. A total of 485 residences were identified. It was assumed that all residences had one self-supplied domestic well. Using the 2000 Census data for Wasco County, the average household was calculated at 2.5 persons (U.S. Census Bureau, 2004). To estimate the water use per capita per day public-supply water use in the city of Mosier was analyzed. Computed water use by city residents was an average of 210 gallons per person per day. City water use tended to be seasonal, using 30 percent of the yearly total from October 15 to April 15 and 70 percent from April 15 to October 15. To calculate self-supplied domestic water use it was assumed that rural residents used water at the same rate as city residents and with the same seasonality. All water use from October 15 to April 15 was assumed to be non-consumptive, returning as groundwater

recharge through septic systems. From April 15 to October 15 it was assumed the same amount (30 percent) was used as in-house use with the balance (40 percent) used consumptively for property irrigation. Defining the coefficient of consumptive use as the percentage of water used consumptively, consumptive water use per well was calculated as:

$$Q = (P)(W)(365 \text{ days})(C), \qquad (D.1)$$

where

P is estimated average household population per residence (2.5 persons),

W is assumed average water use (210 gallons per person per day), and

C is coefficient of consumptive groundwater use (40 percent)

During April 15 through October 15, each well was estimated to use 76,650 gal/yr (approximately 0.24 acre-ft). Consumptive water use was assigned to each well for years starting from the date it was drilled. Where no drill date is available, the well was assigned a construction date from the average date of neighboring wells. The aggregate self-supplied domestic water use for 2006 was estimated at 114 acre-ft (fig. 13).

Appendix E. Details of Groundwater-Flow Simulation

The MODFLOW numerical finite-difference model is an implementation of the conceptual model of the system. This appendix covers the additional technical details of the groundwater-flow simulation completed for this study, and the major topics covered are: (1) the method of representing each of the major flow features; (2) formulation of the model problem for each of the scenarios examined; (3) constraints placed on each model formulation (observations, parameterization, and regularization); (4) model scenario results; and (5) limitations of the model(s).

E.1—Model Grid Design

The model area was discretized into 500-ft sided square grid cells of variable thicknesses, resulting in 100 rows, 120 columns, and 14 layers. Temporal discretization was annual stress periods for the fully transient model, but the modified transient analysis simulates conditions during three distinct periods. Groundwater flow was simulated with MODFLOW-2000 using the Layer Property Flow (LPF) package (Harbaugh and others, 2000). Except for the overburden units, each of the fourteen model layers represents one of the groundwater-flow model units (fig. 3). The overburden was zoned laterally (figs. 3–5) to test the likelihood that glaciofluvial deposits in the OWRD management area limited flow rate during aquifer leakage through commingling wells. The resulting model has 168,000 cells of which 66,600 are active.

The final model grid is rotated 38 degrees clockwise to correspond with major structural features that are known to control flow, namely the Rocky Prairie thrust fault, the Columbia Hills anticline, and the Maupin wrench fault (fig. 22). This aligns the grid with the faults bounding the area of principal interest, which contains the OWRD administrative area and most of the study area water supply wells.

E.2—Additional Details for Model Boundary Conditions

Simulated model boundaries are shown in figs. 22 and A9–A22. Simulated boundaries are discussed in the Model Discretization and Boundaries section of the report, but additional details of simulation of faults, streams, and commingling wells are provided here.

Simulation of Faults

The simulated faults are a simplification of mapped faults and all faults are modeled as vertical, so data collected or modeled in close proximity to modeled barriers inherently contains more uncertainty. The overburden is assumed to be easily deformed, and even if faulting has occurred post-deposition, it is assumed that the faults themselves do not impede flow in the overburden, so no horizontal flow barriers are simulated in layers 1 and 2.

Simulated faults are continuous and span the entire model area, even if offset is small. In areas where faults have small vertical offset, the role of the fault in impeding flow is possibly small. To test whether faults impede flow to varying degrees based on offset or style of faulting, simulated faults were divided into sections (fig. 22). For example, fault sections 31, 32, and 33 represent a gradation from relatively small offset to much larger offset. In this case, section 33 is expected to have relatively lower hydraulic conductance across the fault than section 31.

The simulated faults can also exhibit different hydraulic conductance that varies with depth. For example, at the Rocky Prairie thrust fault, the overthrust thickness corresponds to detachment at the Selah interbed, indicating that older aquifers may still be continuous. As a result, the simulated faults were divided vertically into six groups that allowed testing of fault conductance with depth. Each group contains only one basalt aquifer because these are the preferential groundwater flow paths. The numbering scheme for fault hydraulic characteristic (MODFLOW parameter controlling fault hydraulic conductance) for each fault section is annotated using a single string beginning with "hf" followed by upper layer, lower layer, and fault section in plan view. For example, hf30422 is horizontal flow barrier section impeding flow through layers 3 and 4 along section 22. This provided 48 fault sections that allowed testing of the importance of faults in controlling groundwater flow during the calibration process. Faults are known to be highly important in this flow system (Newcomb, 1969, Lite and Grondin, 1988) and this was the largest set of independent parameters tested with this model. Initial values and regularization constraints are discussed in the parameter estimation section below.

Simulation of Streams

Streams were modeled using a combination of the stream and drain packages (Prudic, 1989; Harbaugh and others, 2000). For the incised streams in the study area, streams flow across large expanses of impermeable CRBG lava flow interior rock, intersecting the thin aquifers relatively infrequently. This provides little opportunity for direct stream loss to the aquifer, but many opportunities for springs and seeps to contribute to streamflow. In model cells containing streams, this preferential gain of streamflow was modeled by setting the drain elevation to the stream stage, resulting in the following formulation of streamflow loss to the aquifer system ($Q_{total\_leakage}$):

$$Q_{total\_leakage} = (C_{stream} + C_{drain})(H_{stream} - H_{aquifer})$$

for

$$H_{stream} < H_{aquifer}, \qquad (E.1)$$

$$Q_{total\_leakage} = C_{stream}(H_{stream} - H_{aquifer})$$

for

$$Z \le H_{aquifer} \le H_{stream}, \qquad (E.2)$$

$$Q_{total\_leakage} = C_{stream}(H_{stream} - Z)$$

for

$$Z > H_{aquifer}, \qquad (E.3)$$

where

Z is the elevation of the stream bottom,

$H_{aquifer}$ is the hydraulic head in the aquifer,

H_{stream} is the stream stage,

C_{stream} is the stream conductane, and

C_{drain} is the drain conductance.

During parameter estimation, the stream and drain conductance were varied as a function of geology. Each conductance can be written as (McDonald and Harbaugh, 1984):

$$C = \frac{KLW}{M}, \qquad (E.4)$$

where

L is the length of the stream or drain,
W is the corresponding width,
M is the thickness across which most of the head loss will occur, and
K is the corresponding hydraulic conductivity.

For each model cell, the length of a stream depends on the path across the cell, and conductance is linearly dependent on the path length. Because this is the only part of the equation that is well known, the dependence on stream length is made explicit, but the other three terms are lumped and treated as a single adjustable parameter during estimation.

In addition to using drains in stream cells, drains were also used at erosional or depositional margins where water may freely drain from hydrogeologic units. For drains occurring in stream model cells, stream geometry also could be used in the parameterization, however the geometry of drains at layer margins is less precisely defined. Rather than treating drains differently at streams and layer margins, a constant length is assumed for all drains, and the drain conductance was varied as a single parameter during parameter estimation.

Because the stream package does not allow water to be routed from drains into the stream, it was necessary to add all drainage and net stream gain to compute $Q_{total\_leakage}$ for comparison with stream-flux calibration targets. This was accomplished by using ZONEBUDGET (Harbaugh, 1990) where each zone is defined to include all cells that can drain to a stream above the associated stream flux target.

Simulation of Commingling Wells

Initially, the multi-node well package (Halford and Hanson, 2002) was used to represent wells in the model, but resulting numerical instability resulted in frequent non-convergence and significant hydraulic budget errors, making use of this package impractical. The chief cause of instability for this model is that the multi-node well (MNW) package solves the groundwater flow equations and intra-borehole fluxes iteratively. Oscillatory behavior of the flow equation solver resulted because the model cell size is small, the hydraulic conductivity contrasts are large, and the storage terms are small. In the event of solver convergence problems, Halford and Hanson (2002) recommend modeling commingling of wells by varying the vertical conductance of cells that contain commingling wells, and modeling pumping using the standard well package (Harbaugh and others, 2000). This has the net effect of moving all commingling and pumping effects into the main MODFLOW equations, rather than requiring iterative solution.

This fix was implemented, and model stability and robustness were greatly improved, allowing investigation of the full range of commingling effects. However, two limitations were imposed on the model analysis by this choice. First, a full transient analysis became impractical, because commingling wells were installed gradually over many years, and hydraulic conductivity is not a time-varying parameter in MODFLOW. As a result, a modified transient analysis was used. Second, the MNW package allows water to be supplied by each cross-connected aquifer as a function of pumping stress on the commingled well, whereas the standard well package requires that pumping stress be applied to each layer individually. To distribute this pumping, it was assumed that the amount of water supplied from each layer was proportional to the fraction of the total transmissivity represented by each layer. Mathematically, the fraction of pumping taken from layer j is:

$$Q_j = Q_{well} \frac{K_j b_j}{\sum_{n=1}^{N} K_n b_n}, \qquad (E.5)$$

where

Q_{well} is the pumping rate of the well,

K is the horizontal conductivity of the layer,

b is the thickness of the layer at the well location (as represented by the thickness of the model cell), and

N is the total number of layers across which the well is screened.

The product Kb is the transmissivity. This transmissivity-weighted average ensures most of the water is from permeable units and the sum of the pumping from the N layers is the total pumping required. Because the conductivities of the layers were varied during parameter estimation, the volume pumped from each layer also was varied.

Assignment of model layers for high capacity wells and deep wells installed in the early 1970s was accomplished using well logs, the digital geologic model, and best professional judgment. An automated method was used to assign the model layers for all other wells. This method applied the following rules to the hydrogeologic framework model:

1. The uppermost aquifer penetrated by the well is the uppermost commingled model layer.

2. If the elevation of the well bottom is known, this elevation is used to estimate the deepest aquifer tapped. In the simple case, this aquifer is the lowermost aquifer that the well penetrated. However, if the well terminates in an overlying confining unit but penetrates 75–100 percent of its thickness, it is assumed that the underlying aquifer is hydraulically connected to the well, because wells generally terminate when a productive aquifer is located.

3. If the elevation of the well bottom is not known, but there are other wells in the quarter-quarter section for which the well bottom elevations are known, the median value is used to estimate the deepest aquifer tapped. If there are no known values, it is assumed that only the uppermost aquifer is tapped. This assumes that for domestic use, drilling will be terminated at the shallowest aquifer.

E.3—Hydraulic Conductivity and Storage Coefficients

There are 16 hydraulic conductivity zones defined in the model. Fifteen of these zones represent hydrogeologic units, with the remaining zone representing pseudo-cells. These pseudo-cells are used to facilitate flow connection between model layers where an intervening layer has pinched out, and therefore is not present. They are 1-ft thick cells with high vertical conductivity and low horizontal conductivity, simulating direct vertical hydraulic connection between the model layers. Use of these pseudo-cells allows the representation of each model layer as a distinct hydrogeologic unit.

Model layers 3–14 are each represented by one hydraulic conductivity zone each, totaling 12 zones. Where glaciofluvial deposits exist, both model layers one and two belong to a single zone. Otherwise, layer one is used to represent the poorly-sorted, relatively low-permeability upper part of the Chenoweth formation, and layer two is used to represent the higher permeability aquifer reported to occur in some parts of the Chenoweth Formation. In every instance, highly permeable model layers are adjacent to low permeability layers. This geometry ensures that vertical groundwater flow will be controlled by low permeability units and horizontal groundwater flow will be controlled by highly permeable units. For the modified transient analysis of commingling wells, all 16 zones were modeled with isotropic hydraulic conductivity, and for the model used to examine management scenarios, the Upper Undifferentiated Overburden (fig. 3) was modeled as anisotropic. The change for the Upper Undifferentiated Overburden was made to address limitations of the model for assessing the value of management scenarios (see Separation of Pumping and Commingling Effects).

The vertical hydraulic conductivity of commingling wells was defined using 16 additional zones, one corresponding to each of the 16 hydrogeologic unit zones previously defined. For each of these additional zones, horizontal hydraulic conductivity was tied to the horizontal hydraulic conductivity of the corresponding hydrogeologic unit zone, and vertical hydraulic conductivity was allowed to represent the effective hydraulic conductivity of the commingling wells. The vertical hydraulic conductivity was assumed to be the same constant value for every well cell, resulting in a single parameter to be investigated, even though vertical well hydraulic conductivity is a function of well diameter and hydraulic gradient between aquifers. It is a reasonable assumption that only one well exists in each 500-ft grid cell, and although it would be practical to compute the effect of well radius, it was not practical to adjust hydraulic conductivity as a function of hydraulic gradient. The effect of well diameter and hydraulic head difference are discussed below when establishing acceptable parameter ranges for the calibration process (see Expected and Calibrated Commingling Well Conductivity, appendix E.5).

Storage coefficients were assigned in the same manner as hydraulic conductivity. Each of the 15 hydrogeologic units and the single pseudo-cell zone has a specific storage and specific yield defined, totaling 32 zones that are coincident with the hydraulic conductivity zones.

E.4—Model Implementation

The parameterization of the model provides a flexible formulation of the conceptual problem that can be used to test the influence of any of the flow features. The model was calibrated, then it was used as a predictive tool in various ways. To systematically explore the model using computer assisted methods (Hill and Tiedeman, 2007), the model independent parameter estimation and prediction software, PEST (Doherty, 2005, 2010), was used. As implemented, PEST required three main groups of information: parameters, observations, and prior information. The parameters are described in the section Groundwater-Flow Model Analyses,

with additional detail provided in appendix sections E.2 and E.3. Observations used include groundwater levels and estimates of groundwater contributions to streamflow (summarized in the Calibration section). Parameters and prior information together define the model parameterization. Prior information describes *a-priori* estimates of parameters and relations between parameters that can be described in equations.

Selection of Parameters to Estimate

The parameters to be tested are summarized in table E1 and the Tikhonov regularization (Doherty, 2005) conditions between parameters are summarized in table E2. As defined, the maximum number of adjustable parameters to be investigated was initially 115 for transient scenarios and 100 for steady-state scenarios. The number of adjustable parameters was further reduced following preliminary model calibration.

Table E1.   Summary of parameter groups used in the groundwater model.

Parameter group	Physical relevance
Hydraulic conductivity of each of the 15 hydrogeologic units.	The rate at which water may be transmitted through the associated unit.
Vertical hydraulic conductivity of commingling well cells.	This represents the rate at which water may be transmitted through the well borehole from one aquifer to another.
Horizontal flow barrier conductance (hydraulic characteristic value) for each of the 48 fault segments.	This represents how easily water may pass through a fault. If conductance is high relative to hydrogeologic unit that it cuts, then the fault does not impede flow; but if conductance is low enough to impede flow, the fault is important in controlling flow.
Drain conductance for each of the 15 hydrogeologic units	Each unit has a separate drain parameter, because the ease with which water drains is assumed to be related to the unit's hydraulic properties.
Stream conductance *fraction* for each of the 15 hydrogeologic units.	It is assumed that the principle control on how easily water is gained or lost to streams is controlled by which hydrogeologic unit it is in hydraulic connection with. An initial best guess of stream conductance was made using stream geometry and an estimate of stream bed conductivity. The stream conductance fraction is a multiplier for each hydrogeologic unit that modifies the cell-by-cell initial best guess of stream conductance.
Recharge *fraction* dependent upon which layer is encountered as the uppermost model layer. There are six parameters, one for each major aquifer that occurs at the land surface.	This allows testing the assumption that recharge to any given unit occurs as predicted by the PRMS model. There are six groups, one for each major aquifer except the base of the Pomona Basalt (never occurs at land surface). The six groups are: (1) Overburden, (2) Pomona flow top and interior, (3) Selah Interbed and the upper Priest Rapids flow top and interior, (4) lower Priest Rapids flow top and interior, (5) lower interbed and Frenchman Springs flow top and interior, and (6) Grande Ronde flow top. The recharge fraction is the multiplier for the array defined by PRMS.
Specific storage terms for each of the 15 hydrogeologic units.	These represent how much water each unit stores and how water is released under the assumption each aquifer is confined. Confined versus unconfined assumptions are discussed in detail in the text.

Table E2. Physical interpretation of prior information specified in the groundwater-flow model.

[**Abbreviations:** PEST, parameter estimation software; MODFLOW, Modular finite-difference flow model; ft/d, foot per day]

Prior information statement	Physical interpretation	Relative weight
The drain conductance of the Pomona flow bottom and both Priest Rapids flow top aquifers is the same.	Because these are all highly permeable basalt interflow zones, it is expected that water will drain similarly from each of these.	Low
The hydraulic conductivity of each layer is the same as the next upper or lower layer of similar morphology.	If no other information is available, the best guess for the permeability of any layer is the permeability of the layer that was deposited under the most similar conditions. Further, if there is a trend in conductivity, it will likely occur with depth due to compaction and chemical evolution of the lava flows.	High if conditions were very similar, but low if the uncertainty is high.
The hydraulic conductivity of the Pomona flow top is ten times higher than for the glacio-fluvial deposits.	It is assumed that the glacio-fluvial deposits would likely impede water flowing freely from the Pomona flow top.	Low.
The horizontal flow barrier conductance of any section is the same as the conductance of the next section above or below it.	The best estimate of how easily a fault transmits water through one layer is the ease with which it transmits water in the vertically adjacent section. Further if there is a trend in conductance, it will likely occur with depth as a result of confining pressure and chemical evolution of the lava flows.	High.
The hydraulic conductivity of the lower Dalles unit is 1,000 times greater than the upper Dalles unit horizontal conductivity.	The bottom of the Dalles unit has been documented as permeable and productive in portions of the study area, but the upper portion is much finer textured. The value was selected to give an effective horizontal to vertical conductivity ratio of 100:1, which is typical for many heterogeneous systems.	Very low.
The hydraulic conductivity of the lower Priest Rapids flow top is ~1,250 ft/d.	This estimate was computed using pump test results from Lite and Grondin (1988), and the thickness of the most likely unit from the geomodel.	Low. Since the value is a very localized sample of a very permeable portion of one of the interflows, it is uncertain if this value is representative of conductivity controlling the watershed scale flow.
The hydraulic conductivity of the Pomona flow bottom is ~2,500 ft/d.	This estimate was computed using pump test results from Lite and Grondin (1988), and the thickness of the most likely unit from the geomodel.	Extremely low. In addition to the caveats immediately above, the pump test location is very near a pinch-out of the unit, making the estimate even more uncertain.
The stream conductance fractions are 0.1.	This is a mathematical trick to aid in mathematical stability both for PEST and the MODFLOW model. Since all stream cells also contain drain cells, loss from the system may equally be achieved by increasing either drain or stream conductance. For PEST, this clarifies which parameter to adjust. For MODFLOW, high conductance of streams sometimes gives stability problems, so whenever possible, stream conductance will be minimized in favor of increases in drain conductance.	Low.

Table E2. Physical interpretation of prior information specified in the groundwater-flow model.—Continued

[**Abbreviations:** PEST, parameter estimation software; MODFLOW, Modular finite-difference flow model; ft/d, foot per day]

Prior information statement	Physical interpretation	Relative weight
Drain conductances for aquifers and the upper Dalles layer are ten times greater than hydraulic conductivities for the same hydrogeologic units.	This is purely for mathematical stability of the estimation process. This prior information prevents drain conductances from becoming arbitrarily high if these parameters become insensitive.	Low.
The horizontal hydraulic conductivity for the upper Dalles zone is 100 times greater than the vertical conductivity.	This condition was only used for generation of vulnerability maps (see Evaluation of Potential Management Options) to prevent anomalously high head values in model panel 1 from skewing results. It is consistent with typical values of anisotropy.	Low.

During model runs where recharge was also adjustable, the general behavior of the model was to shunt water to the observation-data-poor Grande Ronde aquifer when recharge was increased or to reduce the flow to the Grande Ronde when recharge was decreased. The net effect was to take excess water and shed it to the Columbia River through the lowermost aquifer. For this reason, variation of the recharge provided little insight into the governing groundwater-flow processes in the area of interest. Because recharge was estimated using an independent method, and because PRMS recharge values provide a reasonable and conservative estimate, recharge was not adjusted for most of the groundwater flow simulation model analysis. This assumption was relaxed and examined following the bulk of the analysis below.

The transient analysis was limited, so the parameterization of storage terms was never refined. Initially, it was assumed that all sedimentary units had the same specific storage coefficients and all basalt units had the same specific storage coefficients, reducing the number of free parameters from fifteen to two. Simulation runtimes for the preliminary transient model using annual stress periods were on the order of hours, with non-convergence and significant mass conservation errors for some combinations of parameters. Three different layer assumptions were evaluated: (1) Layer 1 unconfined, and all other layers confined; (2) Layer 1 unconfined, and all other layers convertible; and (3) all layers confined. Even though some simulations with an unconfined layer 1 converged faster, the general convergence properties of the model were improved by modeling all layers as confined.

Steady-state simulations were far less time-consuming and all storage terms dropped out of the mathematical formulation, greatly increasing the efficiency of the parameter estimation process. Steady-state simulations converged in approximately 10 seconds per run. To capitalize on the favorable runtimes and robust nature of the steady-state simulations, the problem was reformulated into a modified transient analysis, assuming the system is in a dynamic steady state at three distinct periods. Because this formulation is insensitive to the formulation of storage terms, and because most aquifers are confined, all model layers were simulated as confined to improve convergence during automated parameter estimation.

The final number of independently adjustable parameters for the modified transient analysis was reduced to 85 by fixing the 6 recharge parameters and tying 9 insensitive parameters to sensitive parameters in adjacent hydrogeologic units (7 insensitive drain parameters for confining units were tied to the adjacent aquifer drain parameters, the insensitive Glaciofluvial Aquifer stream parameter was tied to the Undifferentiated Overburden stream parameters, and the poorly constrained hydraulic conductivity of the Grande Ronde flow-top aquifer hydrogeologic unit was tied to the Frenchman Springs aquifer hydrogeologic unit). Numerical stability and improved convergence were accomplished by adding regularization constraints (table E2) using prior information. Weights were only high for two sets of prior information, with low weights generally reserved for prior information that was added to guide the estimation process only when mathematical expediency contradicted physical reasonableness. The high weight sets belong to prior information equations associated with hydraulic conductivity of model layers or conductance of horizontal flow barriers. In both cases, the equations merely state that the flow properties of similar units should be similar, preventing the model from achieving a good fit by giving different values to features that should behave similarly.

Details of the Predictive Uncertainty Assessment of Pumping Compared With Effects of Commingling

Establishing Confidence Intervals for Predictions

Uncertainty in model predictions was evaluated by finding sets of reasonable parameters for which the influence of commingling wells was minimized and maximized. The best fit calibrated model demonstrated that the dominant cause of declines (approximately 85 percent) could be the result of commingling (fig. 26), so it remained to find a set of reasonable parameters that fit the data almost as well, but for which commingling was minimized. A precise definition of "almost as well" was provided by using the Scheffe statistic for simultaneous estimation of parameters with non-linear confidence intervals (Hill and Tiedeman, 2007, p. 177–181). In particular, acceptable error [in terms of the weighted least-squares objective function (ϕ)] was defined using $\hat{\delta}$, corresponding to a confidence interval of greater than 95 percent that satisfies:

$$\phi \le \phi_{min} + \hat{\delta}, \qquad (E.6)$$

where

ϕ_{min} is estimated as the value of ϕ from the best-fit calibrated model.

When computing simultaneous non-linear predictions, $\hat{\delta}$ may be estimated as (Doherty, 2005; Hill and Tiedeman, 2007, p. 178):

$$\hat{\delta} = NP \bullet s^2 \bullet F_\alpha (NP, NDF), \qquad (E.7)$$

where

NP is the number of parameters,

s^2 is the calculated error variance,

NDF is the number of degrees of freedom, and

F_α is the F-distribution with confidence $(1-\alpha)$.

In our case, NFD equals the number of observations, plus the number of prior information equations, minus the number of parameters. Doherty (2005) provides the following estimate for the error variance:

$$s^2 = \frac{\phi_{min}}{NDF}. \qquad (E.8)$$

Substituting (eq. E.8) into (eq. E.7), equation E.6 can be rewritten in terms of the allowable misfit between simulated and observed values by using estimates of the minimum weighted least-squares objective function, number of parameters, the degrees of freedom, and the desired confidence level:

$$\phi \le \phi_{min} \left(1 + \frac{NP}{NDF} F_\alpha (NP, NDF)\right) = \phi_{min}(1+\theta), \quad (E.9)$$

where

θ is defined by equation E.9.

A value for θ of 2.25 corresponds to greater than 95-percent confidence for 63 observations with non-zero weight, 81 prior information statements with non-zero weight, and 85 adjustable parameters.

Predictive Objective Function

A predictive objective function was defined so that PEST could be used to find the minimal commingling well effect resulting from any set of flow model parameters that satisfy the 95-percent confidence criteria defined in the previous section (Establishing Confidence Intervals for Predictions). The effect of commingling wells is minimal if groundwater levels return to pre-development conditions following the cessation of pumping. The recovery of each well was formulated as:

$$head_j^{final} - head_j^{late} = recovery_j, \qquad (E.10)$$

where

$head_j^{final}$ is the value of hydraulic head in well j after pumping is stopped,

$head_j^{late}$ is the value of hydraulic head in well j at late time under pumping conditions.

To measure simulated recovery, total recovery was formulated as:

$$\text{Total Recovery} = \sqrt{\sum_{j=1}^{M} \left(head_j^{final} - head_j^{late}\right)^2}. \quad (E.11)$$

This equation could be used as the predictive objective function except for one potentially significant drawback. When trying to maximize this function to find the set of parameters for which commingling has the minimum effect, the Total Recovery could be dominated by a large recovery in only a few wells. However, figure 9 indicates that Group 1 wells should behave similarly. This potential drawback was addressed by adding the expected recovery based on historical data and a penalty function for when wells behave dissimilarly, yielding the final form of the prediction value to minimize:

$$\psi = \sqrt{\sum_{j=1}^{M}\left(recovery_{estimated}^{final} - \left(head_j^{final} - head_j^{late}\right)\right)^2}$$
$$+\tilde{\lambda}\sqrt{\sum_{j=1}^{M}\left(head_j^{final} - head_{median}^{final}\right)^2}. \qquad (E.12)$$

The argument under the second radical is the penalty function computed as sum of the distances between each final hydraulic head estimate and the median value of all of the final hydraulic heads, ensuring the wells recover together. The $\tilde{\lambda}$ is a weight factor that is manually selected to ensure that the penalty function is non-negligible. This weight was varied to ensure no persistent bias occurred during predictive runs. The $recovery_{estimated}^{final}$ term was added under the first radical so that ψ is a sum of two terms that should both be minimized. All groundwater levels should return to a value between 150 and 175 ft higher, so $recovery_{estimated}^{final}$ was set to 175. The formulation is only sensitive to this value if modeled recoveries approach the selected value, and this did not occur (fig. 26).

Computation of Change in Columbia River Basalt Group Aquifer Storage for Aquifer Vulnerability Mapping and Evaluation of Management Scenarios

For the purposes of computing change in aquifer-system storage to generate vulnerability maps and to assess management options, only change in storage of CRBG basalt aquifers was computed. For each cell, the change in storage was computed as the change in hydraulic head times the area of the cell times a storage coefficient. Total change in storage was computed by summing all model cells representing hydrogeologic units of interest. A single constant value of storage coefficient consistent with specific storage of confined basalt aquifers was used for all CRBG aquifers. This is a limitation of the results, because some aquifers are unconfined and storage change occurs by filling or draining pore spaces rather than by compressing the water and the aquifer material. Future use of a groundwater-flow model with convertible layers may be preferable for some applications. The advantage of the simpler, single storage coefficient approach is that the comparative analysis of high, medium, and low vulnerability areas are independent of the values of storage terms.

E.5—Additional Observations and Limitations from Groundwater-Flow Simulation Results

This section contains an analysis of current limitations of the groundwater-flow model for replicating aquifer-system response to commingling wells over time. These observations may provide guidance for future modeling strategies in Mosier and the larger Columbia River Basalt aquifer system.

Expected and Calibrated Commingling Well Conductivity

Considering the fact that only a couple of commingling wells were installed each year between 1972 and 1976 (fig. 16), it is evident that only a few commingling wells in a vulnerable area may cause significant declines. During the calibration and predictive analyses, a range of parameters were explored, and the following general conclusions may be drawn about the effective vertical hydraulic conductivity of well cells in the model and the hydrogeologic system in general. The estimated vertical hydraulic conductivity of cells with commingling wells ranged from large insensitive values (greater than 10,000 ft/d) to sensitive values that allow the transmission of water at rates similar to basalt aquifer conductivity (about 0.2 ft/d). The vertical hydraulic conductivity of the well cells becomes sensitive in the parameter estimation when it starts to impede vertical flow, and as expected, estimated vertical hydraulic conductivity of well cells was lowest for the maximum predicted recovery scenario (fig. 26).

A-priori estimates of effective vertical hydraulic conductivity of well cells were made to ensure that the calibrated values are reasonable. This was accomplished by computing the vertical hydraulic conductivity of a 500 ft model cell that would provide equivalent Darcian flow as turbulent flow through a vertical borehole as approximated by a rough-walled pipe. Setting the flows equal to each other:

$$Q_{pipe} = Q_{Darcy} = -K_{eff}^v A \frac{\Delta h}{L}, \qquad (E.13)$$

where
- A is the area of the model cell orthogonal to flow,
- Δh is the hydraulic head across the cell in the vertical direction,
- L is the length over which the hydraulic head is dropped (the thickness of the cell), and
- K_{eff}^v is the effective vertical hydraulic conductivity for which the estimate was made.

Rearranging yields:

$$K_{eff}^{v} = \frac{-Q_{pipe}L}{A\Delta h}.$$ (E.14)

It remains to estimate flow through a pipe subject to the same hydraulic head gradient. The Navier-Stokes equation can be solved for laminar flow in a smooth pipe (Welty and others, 1969, p. 106–109), yielding estimates of K_{eff}^{v} ranging from about 5,000 to about 25,000 ft/d for wells ranging in diameter from 8 to 12 in. diameter, respectively. However, when considering the magnitude of pre-development gradients across basalt flow interiors, turbulent flow is likely to occur in boreholes, particularly in early time after wells were installed. Historically, 70–100 ft of hydraulic head difference occurred between aquifers separated by a hundred or more feet of impermeable basalt (Lite and Grondin, 1988). Considering that the laminar flow approximation provides an upper bound, the fully turbulent flow case provides a lower bound to K_{eff}^{v}. Turbulent flow in a rough pipe can be described by the following relations (Welty and others, 1969, p. 194–200):

$$\frac{1}{\sqrt{f_f}} = 4.0\log_{10}\left(\frac{D}{e}\right) + 2.28,$$ (E.15)

where

D is pipe diameter,
e is pipe roughness (units of length), and
f_f is the Fanning friction factor defined by the relation:

$$h_L = 2f_f\frac{L}{D}v^2,$$ (E.16)

where

h_L is the hydraulic head loss expressed in units of $\frac{\Delta P}{\rho}$,

where

ΔP is the pressure gradient,
ρ is the water density, and
v is the fluid velocity.

For a hydraulic head gradient in the horizontal direction in groundwater hydrology, $\Delta h = \frac{\Delta P}{\rho g}$, allows conversion of h_L to the same notation as the Darcy formulation (eq. E.13). Q_{pipe} is this velocity times the area of the pipe, providing all relations necessary to compute K_{eff}^{v} for fully turbulent flow in rough pipes:

$$K_{eff}^{v} = \frac{\pi\sqrt{g}\left(\frac{D}{2}\right)^{\frac{5}{2}}}{A\sqrt{\frac{\Delta h}{L}}}\left[4.0\log_{10}\left(\frac{D}{e}\right) + 2.28\right].$$ (E.17)

All parameters in this equation are well known, except pipe roughness, so that K_{eff}^{v} can be plotted as a function of hydraulic head gradient. The pipe roughness term has units of length and can be conceptualized as a characteristic height of projections from the pipe wall (Welty, Wicks, and Wilson, 1969). Riveted steel or concrete pipes are rougher than most pipes, with a roughness typically ranging from 0.0002 to 0.002 ft, so a somewhat conservative estimate of 0.02 ft was used to produce figure E1. This figure illustrates that the range of calibrated effective vertical conductivities is reasonable, but it also illustrates limitations of the model.

First, the parameterization assumes that all vertical well cell hydraulic conductivities are the same, but figure E1 shows that to the contrary, this is a function of gradient, and by induction, position in the watershed. For this reason, model fit can be worse for late-time simulations with a significant number of spatially diverse commingling wells that use a single value of vertical hydraulic conductivity.

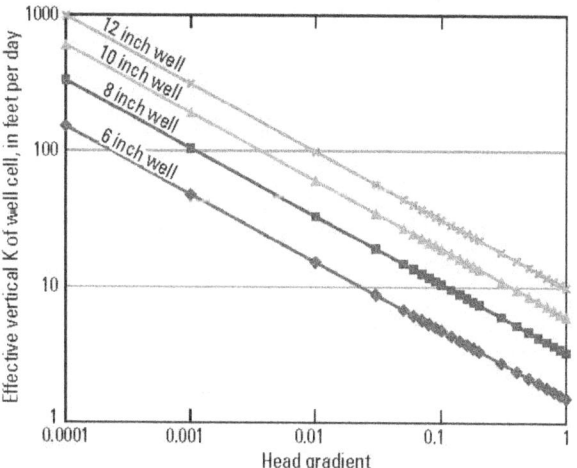

Figure E1. *A-priori* estimates of effective vertical hydraulic conductivity of well cells as a function of hydraulic head gradient. Conductance varies as a result of turbulent losses along the well borehole.

Second, the *a-priori* estimated values of vertical well cell hydraulic conductivity are in the range where sensitivity to this parameter decreases rapidly (compare the two values and sensitivities in fig. 24). The value of 0.1 ft/d (fig. 24) is approximately where flow through the well starts to become limiting, which explains why the parameter is sensitive at this value. In this context, the *a-priori* estimates (fig. E1) indicate that the wellbore itself is not likely to be flow limiting, but rather, turbulent losses within the formation as flow radially converges at the borehole may be more restrictive. Assuming that flow paths within the formation also can be represented as flow through rough pipes, equation E.17 may be used to estimate the effective diameter of pores controlling flow. The analysis was crude, but indicated that characteristic pore diameters of the rate-controlling aquifer may range from 1 to 4 in., corresponding to approximately 0.1 ft/d. Considering basalt aquifer morphology, these values appear to be reasonable. The calibrated value of well cell vertical hydraulic conductivity may be accounting for hydraulic head loss near the borehole, and not in the borehole itself.

Third, the strong dependence of the effective vertical hydraulic conductivity on the hydraulic head gradient coupled with the hydraulic head loss over time indicates any model with a constant well vertical hydraulic conductivity possibly will have severe limitations for use in a fully transient analysis. Because effective well vertical hydraulic conductivity increases as the head gradient decreases (fig. 9), the resulting drawdown curve will be flatter over time than for the response resulting from having a constant well vertical hydraulic conductivity, potentially contributing to the linear system response of the Group 1 groundwater level declines (fig. 9).

Observations from Fully Implemented Transient Simulations

During preliminary analyses, transient groundwater flow and changes in groundwater levels were simulated using monthly and annual stress periods, generally resulting in convergence problems and poor mass balances. Following the modified transient analysis used for calibration and prediction, a fully transient model was used to investigate the rate of decline observed in the OWRD administrative area. This model used the simulated head from the early-time steady-state model as the initial hydraulic head distribution, and used the late-time distribution of commingling wells to simulate time-varying system response. Because declines have been linear since the 1970s, the late-time (2006) well distribution was assumed to be sufficient for testing the model.

This modeling was accomplished to evaluate the linear nature of groundwater-level declines of CRBG aquifers in the OWRD administrative area. If a single penetration is made into a confined aquifer with a fixed elevation controlling the rate of drainage, then the time-dependent groundwater level decline is predicted to be exponential. Adding penetrations sequentially would increase the rate of decline over time, which may contribute to a more linear appearance, but would more likely result in a variable rate of decline with jumps in rate when new penetrations occur. However, measured groundwater-level time series are persistently linear, and examination of new well hydrographs show that groundwater level in the well will often drop from an initially higher value to approximately the same hydraulic head as the remainder of the group. This argues against sequential well installation as being the primary cause of the linear response, and the absence of changes in the rate of decline associated with new wells implies that the final distribution of wells is sufficient for testing the transient response of the groundwater model.

Recall that all model runs simulated confined groundwater flow in each model layer. The value of the specific storage was adjustable during automated parameter estimation to account for the drainage (specific yield) of some areas of the formations. However, only two specific storage parameters were used initially: one for all basalts and one for all sedimentary units. Early attempts at monthly simulation showed flashy system response of simulated hydraulic heads due to seasonal variations in recharge and variations between years. To reduce this effect, annual average recharge was used with annual time steps for the transient model. This is reasonable because groundwater levels indicate that seasonal and intra-annual effects are small compared to the large declines being analyzed (see the "Temporal Variation in Groundwater Levels and Changes in Groundwater Storage" section of the main report).

Because it was assumed that pre-development and early-time conditions were essentially in steady state, calibration targets for these time periods were used in the same manner as described for the modified transient analysis. Following each early-time steady-state model run, final hydraulic heads were exported to a transient model for use as initial hydraulic heads for the annual time step transient simulations, which allowed examination of system response since the early-1970s to current pumping and commingling stresses. All available hydraulic head measurement values were used. If multiple measurements were taken throughout the year, the median value was used as the annual groundwater level calibration target to de-emphasize outliers resulting from pumping conditions. Under the previous assumptions, parameter estimation using PEST was undertaken. Because groundwater-level measurements were taken at regular intervals during the entire period 1972–2006, the automated calibration was anticipated to find a set of parameters resulting in linear declines of Group 1 wells, as well as matching pre-development and early-time heads.

Using *a-priori* estimates of basalt aquifer storage coefficients resulted in rapid exponential groundwater-level declines of water levels in Group 1 wells, with the system asymptotically approaching steady state in 2–5 years. This timescale of response is similar to the time it took for water levels in several new Group 1 wells to decline from their initial value (shortly after drilling) to values similar to other Group 1 wells (fig. 9), suggesting these wells are located in confined aquifers that were connected during well construction to a portion of the groundwater system experiencing linear declines.

During calibration, storage terms increased from true confined storage values and drainage and commingling parameters slowed water flow from the aquifer system, resulting in a best fit that exhibited an exponential rate of decline with systematic under prediction of groundwater levels in earlier time and over prediction in later time. As a result, model fit was poor, and declines were nonlinear. A suite of runs using different starting values of parameters was explored to ensure the calibration problems were not the result of poor starting values, but in all cases, the model had similar behavior. In other words, the linear declines were not reproduced using the simple two-storage coefficient (overburden and basalt) representation.

The following mechanisms are not represented in the model, and some combination of these may account for the approximately linear response of the system:

Non-Darcian flow resulting from commingling wells: Turbulent hydraulic head loss will result in lower apparent hydraulic conductivity under high gradient conditions, with well vertical hydraulic conductivity apparently increasing as hydraulic heads between aquifers equilibrate. This mechanism was discussed more completely in the previous section.

Dual storage parameters representing specific yield as well as specific storage: In a single-aquifer groundwater-flow system, this does not have a large effect, but in a multi-aquifer flow system, drainage of the one aquifer (specific yield) through commingling wells into a second confined aquifer can result in complex behavior.

The Effect of Compartmentalization on Transient Behavior

To test the efficacy of the dual storage mechanism for linearizing declines in a compartmentalized system, a simple two compartment analytical model was developed (fig. E2). Initially, the system is composed of two isolated compartments at different initial steady hydraulic heads. At time zero, each compartment is perforated, resulting in two effective conductance terms that describe how water flows between the compartments and out of the system as a function of the difference in hydraulic head across the perforated barrier. Perforation of the compartments connects compartment 1 with compartment 2 and compartment 2 with a fixed hydraulic head condition outside the system. It is assumed hydraulic head in compartment 1 is greater than hydraulic head in compartment 2, which is in turn, greater than hydraulic head outside the system.

This geometry is a simplified representation of the case where deep basalt aquifers are at a higher hydraulic head than shallow basalt aquifers, and where commingling wells

create a conduit to a constant elevation outflow such as what is presumed to exist near Group 1 wells. It is a reasonable representation for the case where flow between the aquifers is more restrictive than flow through the aquifers themselves. The method of solving the following differential equations guarantees that the general solution may be written as the sum of the solutions to the homogeneous equations and the particular solutions (Powers, 1987). The solution to the homogeneous equations defines the transient response of the system with final steady-state heads defined by the particular solution, which is determined by the recharge rate and the conductance out of each compartment. Because we are interested in examining the time-varying response of the system, it is sufficient and simplest to examine the solution to the homogeneous equations. The solution to the homogeneous equation for each compartment is a good approximation to the general solution for the case where the leakage rate is much greater than the recharge rate.

The equation describing hydraulic head in compartment 1 is:

$$\hat{S}_1 \frac{dh_1}{dt} = -C_{1 \leftrightarrow 2}\left(h_1 - h_2\right), \qquad \text{(E.18)}$$

where

$\hat{S}_1$ is a coefficient describing storage of compartment 1,

h_1 and h_2 represent the hydraulic heads in compartments 1 and 2, respectively, and

$C_{1 \leftrightarrow 2}$ is the conductance between compartments 1 and 2.

Similarly, hydraulic head in compartment 2 is described by:

$$\hat{S}_2 \frac{dh_2}{dt} = C_{1 \leftrightarrow 2}\left(h_1 - h_2\right) - C_{2 \leftrightarrow f}\left(h_2 - h_f\right). \quad \text{(E.19)}$$

The final hydraulic head to which the system will equilibrate (h_f) is controlled by the elevation of the outfall and can be set to any arbitrary datum defining the elevation from which all other hydraulic heads are measured. For this analysis, it is set to zero, simplifying equation (E.19) to:

$$\hat{S}_2 \frac{dh_2}{dt} = C_{1 \leftrightarrow 2}\left(h_1 - h_2\right) - C_{2 \leftrightarrow f} h_2. \qquad \text{(E.20)}$$

The conductance across each barrier is the change in volumetric flux per unit change in hydraulic head across the barrier, giving units of length squared per time. The coefficient describing storage of each compartment is equal to the volume of water released per unit change in hydraulic head, or in hydrogeologic terms:

Two-compartment Model

Initial Condition = Two sealed compartments

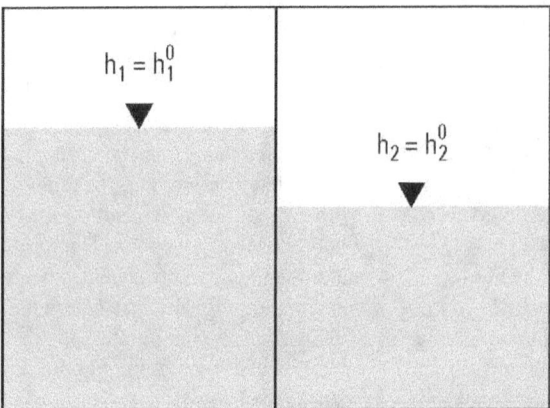

Transient Condition = Both compartments have been perforated

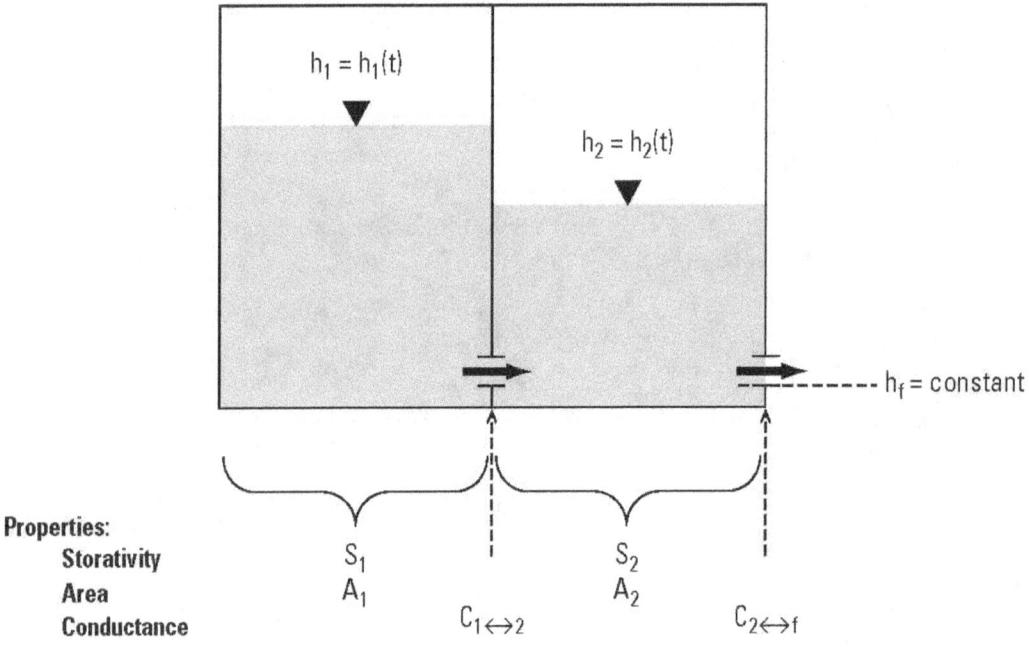

Figure E2.   A two-compartment model.

$$\hat{S}_k = A_k S_k, \tag{E.21}$$

where

A_k is the plan view area of compartment k, and
S_k is the storativity (the usual storage coefficient used in groundwater flow equations) of compartmen k.

Writing the equations (eq. E.18) and (eq. E.20) in terms of one independent variable each yields:

$$S_1 A_1 S_2 A_2 \frac{d^2 h_1}{dt^2} + \left[\left(C_{1\leftrightarrow2} + C_{2\leftrightarrow f} \right) S_1 A_1 + C_{1\leftrightarrow2} S_2 A_2 \right]$$
$$\frac{dh_1}{dt} + C_{1\leftrightarrow2} C_{2\leftrightarrow f} h_1 = 0, \tag{E.22}$$

and

$$S_1 A_1 S_2 A_2 \frac{d^2 h_2}{dt^2} + \left[\left(C_{1\leftrightarrow2} + C_{2\leftrightarrow f} \right) S_1 A_1 + C_{1\leftrightarrow2} S_2 A_2 \right]$$
$$\frac{dh_2}{dt} + C_{1\leftrightarrow2} C_{2\leftrightarrow f} h_2 = 0. \tag{E.23}$$

If the storativities and conductances are known, both equations are homogeneous second order ordinary differential equations with unique solutions of the form:

$$h_1 = B_1 e^{m_1 t} + B_2 e^{m_2 t}, \tag{E.24}$$

and

$$h_2 = B_3 e^{m_1 t} + B_4 e^{m_2 t}. \tag{E.25}$$

With m_1 and m_2 being the two roots of x for the following equation:

$$S_1 A_1 S_2 A_2 x^2 + \left[\left(C_{1\leftrightarrow2} + C_{2\leftrightarrow f} \right) S_1 A_1 + C_{1\leftrightarrow2} S_2 A_2 \right] x$$
$$+ C_{1\leftrightarrow2} C_{2\leftrightarrow f} = 0. \tag{E.26}$$

Additionally, the B coefficients are provided by solving the initial value problem with two initial conditions for each equation. The initial conditions are that hydraulic head in each compartment is known, and the instantaneous flux honors the following conductance based formulation, which is assumed to hold for all time following perforation:

$$h_1 (t = 0) = h_1^0, \tag{E.27}$$

$$S_1 A_1 \frac{dh_1 (t = 0)}{dt} = -C_{1\leftrightarrow2} \left(h_1^0 - h_2^0 \right), \tag{E.28}$$

$$h_2 (t = 0) = h_2^0, \tag{E.29}$$

$$S_2 A_2 \frac{dh_2 (t = 0)}{dt} = C_{1\leftrightarrow2} \left(h_1^0 - h_2^0 \right) - C_{2\leftrightarrow f} h_2^0. \tag{E.30}$$

This initial boundary value problem was solved for a set of parameters representing the storage properties (area and storativity values representative of specific storage and specific yield) of the Mosier system, allowing assessment of fitted parameters for reasonableness (table E3). All units are in feet and days to allow easy comparison with MODFLOW parameters. The area of compartment 1 is assumed to be a square approximately 6 miles on a side, and compartment 2 is one-half the area corresponding to the approximate size of the lower and upper aquifers respectively. Initial head for compartment 2 was fixed at 300 ft, which corresponds to the approximate head difference between the uppermost aquifers and the elevation of Mosier Creek in the OWRD administrative area prior to 1970. Initial head for compartment 1 and both conductance terms were fitting parameters. Two scenarios were considered for storage terms (table E3): (1) both compartments are assumed to have an equal confined or pseudo-confined value (*equal storage* [table E3; fig. E3]); and (2) compartment 1 is assumed to release water through drainage (specific yield), and compartment 2 has the confined or pseudo-confined value (*unequal storage* [table E3; fig. E3]). Pseudo-confined is defined as a larger than anticipated value for true confined conditions. The specific yield of compartment 1 is fixed at 0.2, and the role of the magnitude of the confined or pseudo-confined storage terms is varied to examine the role of this term on system response (fig. E3).

The starting head in compartment 1 and the conductance terms were varied to achieve an approximately linear decline of 175 ft during a 30 year period, yielding reasonable physical values for both initial head and conductance. Because it is not certain whether water leaks more easily from the commingled basalt system or more easily between the basalt aquifers within the system, both cases were examined by varying conductance in a fixed ratio during exploration of parameter values.

If more commingling wells are present in the geologically higher aquifers, then conductance could be higher between compartment 2 and the outfall than between the compartments. This corresponds to figures E3D through *F*, where conductance to the outfall is assumed to be twice the conductance between the compartments. The other case is where conductance between the compartments is higher than conductance to the outfall. Because all wells have a sanitary seal, flow up and out the boreholes has to pass through the uppermost geologic units. In most cases, it is feasible that these units provide more resistance to flow than is experienced by the borehole itself, so to test the feasibility of this case, conductance of the outfall was assumed to be 60 percent of conductance between the compartments (figs. E3A–C).

Table E3. Values tested to demonstrate that compartmentalization using reasonable parameter values can help explain the long term linear declines of Group 1 wells.

[**Abbreviations:** ft, foot; ft^2/d, foot squared per day; ft^2, square foot]

Parameter	Figure E3A	Figure E3B	Figure E3C	Figure E3D	Figure E3E	Figure E3F
Initial head in compartment 1 (h_1^0; units = ft)	500	500	500	900	900	900
Initial head in compartment 2 (h_2^0; units = ft)	300	300	300	300	300	300
Conductance between compartments 1 and 2 ($C_{1\leftrightarrow2}$; units = ft^2/d)	4.90×10^4	4.90×10^4	4.90×10^4	2.45×10^4	2.45×10^4	2.45×10^4
Conductance between compartment 2 and the system outfall ($C_{2\leftrightarrow f}$; units = ft^2/d)	2.94×10^4	2.94×10^4	2.94×10^4	4.90×10^4	4.90×10^4	4.90×10^4
Storage coefficient of compartment 1 (S_1; unitless) *Equal storage*	1.0×10^{-2}	1.0×10^{-3}	1.0×10^{-6}	1.0×10^{-2}	1.0×10^{-3}	1.0×10^{-6}
Unequal storage	0.20	0.20	0.20	0.20	0.20	0.20
Storage coefficient of compartment 2 (S_2; unitless)	1.0×10^{-2}	1.0×10^{-3}	1.0×10^{-6}	1.0×10^{-2}	1.0×10^{-3}	1.0×10^{-6}
Area of compartment 1 (A_1; units = ft^2)	1.0×10^9	1.0×10^9	1.0×10^9	1.0×10^9	1.0×10^9	1.0×10^9
Area of compartment 2 (A_2; units = ft^2)	5.0×10^8	5.0×10^8	5.0×10^8	5.0×10^8	5.0×10^8	5.0×10^8

The set of values tested (table E3) demonstrates that compartmentalization using reasonable initial conditions and parameter values can help explain the long term linear declines of Group 1 wells for a range of storage parameter contrasts (fig. E3). For both conductance conditions (compare columns of fig. E3), an approximately linear decline of about 175 ft over 30 years occurs in compartment 2 for the unequal storage case. Further, the effect of varying the confined or pseudo-confined storage term has negligible effect on the unequal storage response as long as this parameter is at least 20 times smaller than the specific yield, indicating that the aquifer being drained controls the rate of decline. As the magnitude of the pseudo-confined storage approaches the specific yield, the solution to the equations becomes sensitive to the volume of water in compartment 2.

The range of 500 to 900 ft for initial hydraulic heads in compartment 1 is reasonable given measured hydraulic heads high in the watershed and anecdotal evidence of hydraulic gradients in the OWRD administrative area. The model predicts that these maximum hydraulic head values only need to have existed prior to aquifer cross-connection. Variations in initial hydraulic head for compartment 1 account for short term increases or declines in compartment 2 hydraulic head following cross-connection events. For example, figures E3A through C, show a minor increase in hydraulic head initially, but lowering initial hydraulic head in compartment 1 to 475 ft eradicates the early-time rise with only a small effect on the remainder of the hydrograph.

To evaluate the magnitude of the conductance terms for reasonableness, Darcy's flow law is compared to the conductance formulation. Setting these equal to each other:

$$-C_{1\leftrightarrow2}\left(h_1 - h_2\right) = -K_v^{wells} A \frac{\left(h_1 - h_2\right)}{L}, \qquad (E.31)$$

where

> A is the area through which the commingling flow occurs, and
> L is the thickness of the barrier penetrated by the well.

The area is the number of well cells (n) times the cell area (250,000 ft^2), yielding:

$$K_v^{wells} = \frac{C_{1\leftrightarrow2}L}{250,000n}. \qquad (E.32)$$

Taking the highest value of conductance from table E3, and assuming a reasonable value of L for a typical vertical distance between aquifers (approximately 100 ft), yields:

$$K_v^{wells} \approx \frac{20}{n} \text{ feet per day.} \qquad (E.33)$$

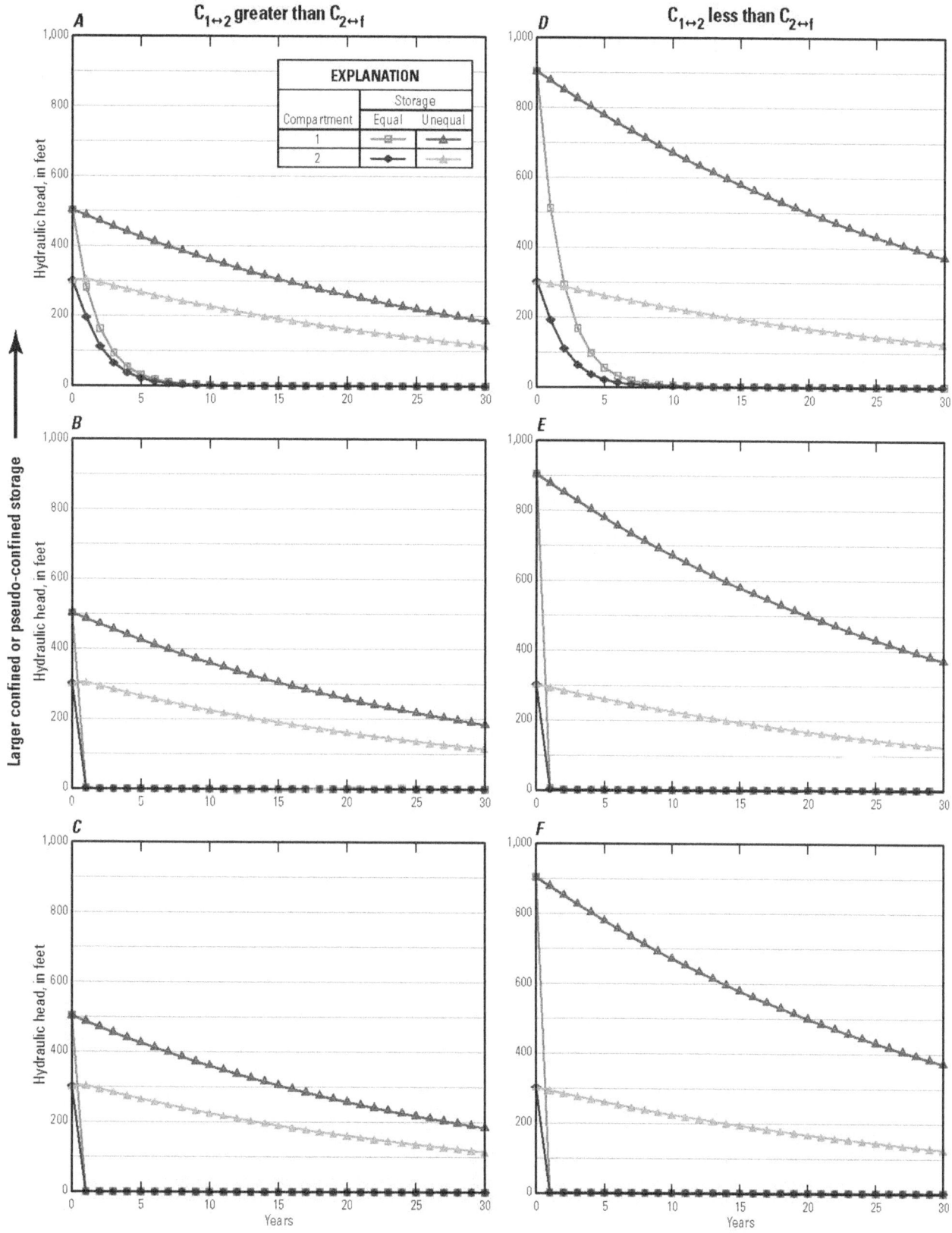

Figure E3.   Hydraulic head response in a two-compartment model under various conditions summarized in <u>table E3</u>.

This implies that one well at 20 ft/d, two wells at 10 ft/d, or 20 wells at 1 ft/d would all provide conductance values consistent with the analytic model. Even one well at 20 ft/d is reasonable given the *a-priori* estimates of conductivity (fig. E1) and falls within the range of calibrated values from the MODFLOW model (fig. 24) and the number of potentially commingling wells (fig. 16). Because both cases considered provided reasonable starting hydraulic head values and conductance values, the model is robust in either case, and both working hypotheses must be retained.

Two final comments on the analytic model are instructive. First, the analytic model is generally insensitive to the size of compartment 2 (data not shown). This is because the source of water controlling the rate of decline is compartment 1. Second, if storativity in compartment 2 also is set to a value of specific yield, then the results are virtually indistinguishable from only using specific yield in compartment 1 (for the conditions analyzed). This again is the result of compartment 1 supplementing compartment 2. However, this begs the question: Why did parameter estimation using the MODFLOW model not drive storage to values consistent with unconfined conditions? This is because there are many more wells besides Group 1 wells for which calibration targets were used. The response of these other wells also places constraints on the model calibration, and the net result is that a single storage parameter is not viable for the entire model area. Use of convertible layers to allow simulation of unconfined conditions allows use of multiple parameters without prior knowledge of which aquifers will drain to supplement other aquifers.

The key conclusions from the analytic model are that compartmentalization and draining of one aquifer to supplement another are viable mechanisms to explain the long-term linear declines of group 1 wells. Additionally, a groundwater-flow model capable of simulating unconfined conditions for all aquifers could be used to test the hypothesis that long-term linear declines are the result of supplementing lower hydraulic head aquifers by draining higher hydraulic head aquifers.

Appendix F. Geophysical Testing of Boreholes

F.1—Borehole Flow Meter Test

Borehole geophysical information was collected in the City of Mosier number 3 well (454033121230101) to identify the depth of permeable productive zones, quantify the contribution from these zones, estimate vertical flow in the borehole between permeable zones, and evaluate the integrity of the well seal and loss of water from the borehole, if any. At the well, groundwater is under pressure and naturally flows from the well unless the well is capped, or shut in. Based on drillers' reports and analysis of cuttings, the well penetrates Pomona basalt, the Selah interbed, Priest Rapids basalt, and enters the top of the Frenchman Springs basalt. Within the Pomona basalt, a potentially permeable zone, identified as broken basalt on the driller report, occurs at a depth of 86 ft (fig. F1). A sedimentary interbed from a depth of 230 to 270 ft separates the Pomona basalt from the underlying Priest Rapids basalt. Based on the driller's description of cracked or broken (fractured) basalt, potentially permeable zones occur at depths of 285, 310, and 368 ft in the Priest Rapids, at 390 and 398 ft at the top of the Frenchman Springs. Although fractured zones of basalt typically are associated with enlargements of the borehole diameter, the borehole caliper log shows little variation in borehole diameter below 275 ft in depth (fig. F1).

Steel casing, with hydraulic seals, extends from land surface to 275 ft in depth, which is designed to isolate the Pomona basalt and the sedimentary interbed from the well. Below a depth of 275 ft, the well is an open borehole, which allows water to enter and leave the borehole to the Priest Rapids and Frenchman Springs basalt and associated interbed.

Imagery of borehole conditions from a video camera confirmed the depths of potentially permeable zones identified as cracked or broken basalt in the drillers' reports. The video images also indicated upward flow of water in the borehole at the base of the casing under shut-in conditions suggesting that water is flowing outside the casing.

Measurements of vertical flow in the borehole with a flowmeter (fig. F1) indicate maximum upward flow at the base of the casing and an abrupt decline in flow in the cased area of the well under shut-in and flowing conditions. The difference in flow below the casing and in the casing is the groundwater flowing upward outside of the casing between the borehole and the casing. Productive permeable zones, characterized by increases in vertical flow in the borehole from 375 to 400 ft in depth and from 300 to 325 ft in depth, represent intervals where groundwater flows from the basalt aquifer system to the borehole. The increase in flow in the borehole is greater from 375 to 400 ft in depth than from 300 to 325 ft in depth, indicating that the deeper interval contributes more water to the borehole.

Fluid temperature and resistivity logs provide information on changes of fluid properties with depth. These changes can be associated with changes in flow into or out of the borehole. Changes in fluid properties at 375 and 310 ft in depth correspond to productive zones in the flowmeter log where water enters the borehole.

Under shut-in conditions, there is no vertical flow in the cased area of the well. Under flowing conditions, water was measured flowing from the well at 55 gal/min. By calibrating the response of the flowmeter to this measured flow, flow in the borehole and flow from the borehole to the annular space can be estimated. Under shut-in conditions, 70 gal/min of upward flow is measured immediately below the casing at 275 ft in depth. Because there is no measured flow in the casing, all of the 70 gal/min of flow exits the borehole at the bottom of the casing. Under flowing conditions, upward flow is 190 gal/min immediately below the casing and 55 gal/min in the casing. The difference between these measurements (135 gal/min) is the net flow leaking outside the cased area of the well. This approach probably represents estimates of maximum flow in the uncased borehole and exiting the borehole.

F.2—Temperature Probe Screening Tool

A miniature pressure and temperature probe was tested as a tool for pre-screening potentially commingling wells for future geophysical logging. Geothermal gradients in the proximal Hood Basin are typically 1–2°C per 100 ft (Grady, 1983), and it was assumed that wells with no commingling would have similar thermal gradients in the well water. If there is significant commingling, then isothermal zones will occur in the region of flow through the borehole.

The City of Mosier number 3 well (454033121230101) exhibits this behavior with temperature varying in zones of groundwater flow contribution to intra-borehole flow, and with relatively constant temperature in zones of constant borehole flow (fig. F1). Even though borehole flow measurements are zero in the well casing during shut-in conditions, flow continues outside the casing, and the isothermal signature persists between approximately 100 and 300 ft. This is the result of thermal conduction into the stagnant water in the well casing from outside the casing where intra-borehole flow was occurring. The temperature log indicates that near 90 ft depth, up-borehole flow is exiting the system, possibly through the "broken grey basalt" recorded on the well construction report between 86 and 87 ft.

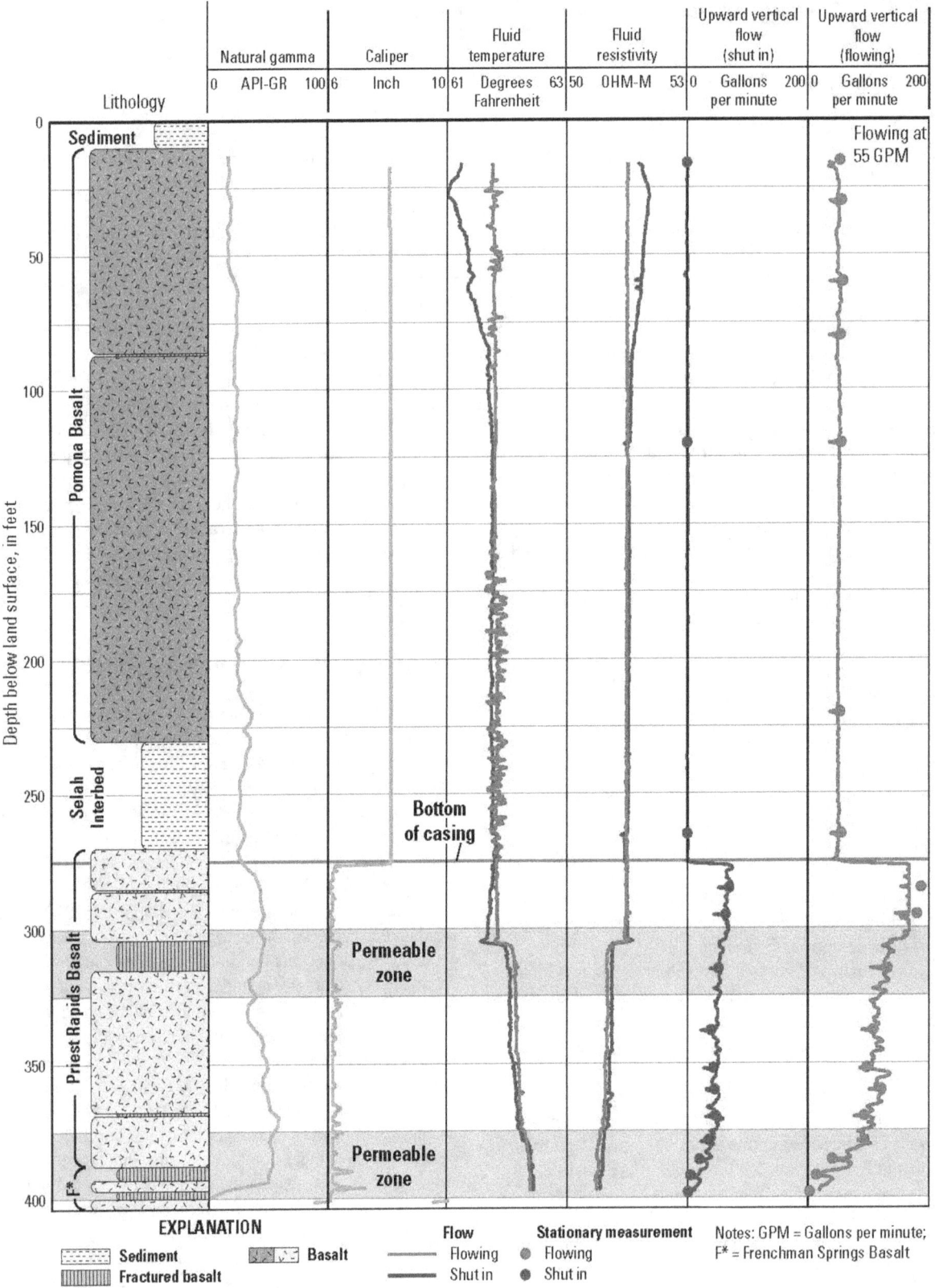

Figure F1.   Results of borehole geophysical logging.

The pressure and temperature probe tested as a screening tool was a Data Storage Tag micro Temperature and Depth (DST micro TD) purchased from Star-Oddi. This device is self contained and measures 8.3 mm by 25.4 mm. The device has a rated maximum depth of 150 m with a resolution of 12 cm. The rated temperature range was -1 to +40 °C with a resolution of 0.032 °C. The probe was attached to a water level probe and lowered into wells past pumps and pump linkages if possible.

The probe was tested on seven wells considered to have a relatively high potential to commingle. Results were mixed with the probe hanging up in three of the wells, although some useful data were collected on one of these wells before getting stuck. The temperature signature indicated potential commingling in four wells, indicating the screening tool is viable for use as a rapid screening tool for wells with sufficient clearance to allow lowering the probe.

www.ingramcontent.com/pod-product-compliance
Lightning Source LLC
Chambersburg PA
CBHW081455170526
45166CB00008B/2436